AF585730

Springer Series in Synergetics

Editor: Hermann Haken

Synergetics, an interdisciplinary field of research, is concerned with the cooperation of individual parts of a system that produces macroscopic spatial, temporal or functional structures. It deals with deterministic as well as stochastic processes.

Volumes 1–39 are listed on the back inside cover

H. Atlan I.R. Cohen (Eds.)

Theories of Immune Networks

With 23 Figures

Springer-Verlag Berlin Heidelberg New York
London Paris Tokyo Hong Kong

Professor Henri Atlan
Department of Medical Biophysics, Hadassah University Hospital, P.O. Box 12000,
IL-51120 Jerusalem, Israel and
Faculté de Médecine, Broussait-Hotel Dieu, Université de Paris VI, 15 rue de l'Ecole de Médecine,
F-75008 Paris, France

Professor Irun R. Cohen
The Weizmann Institute of Science, Rehovot, Israel

Series Editor:

Professor Dr. Dr. h. c. Hermann Haken
Institut für Theoretische Physik der Universität Stuttgart, Pfaffenwaldring 57/IV,
D-7000 Stuttgart 80, Fed. Rep. of Germany and
Center for Complex Systems, Florida Atlantic University,
Boca Raton, FL 33431, USA

ISBN-13: 978-3-642-83937-5 e-ISBN-13: 978-3-642-83935-1
DOI: 10.1007/978-3-642-83935-1

Softcover reprint of the hardcover 1st edition 1989

2154/3150-543210 – Printed on acid-free paper

Preface

For a long time, immunology has been dominated by the idea of a simple linear cause-effect relationship between the exposure to an antigen and the production of specific antibodies against that antigen. Clonal selection was the name of the theory based on this idea and it has provided the main concepts to account for the known features of the immune response.

More recently, immunologists have discovered a wealth of new facts, in the form of different regulatory cells (helpers, suppressors, antigen presenting cells), genetic determinations of immune responses such as those involved in graft rejections, different molecular structures responsible for intercellular interactions such as interleukins, cytokins, idiotype-antiidiotype recognition and others. While furthering our understanding of the local interactions (molecular and cellular) involved in the immune response, these discoveries have led to a questioning of the simplicities of the classical clonal selection theory. It is clear today that every single immune response is a cooperative phenomenon involving several different molecular and cellular interactions taking place in a coupled manner. In addition, cross reactivity to different antigens has shown that responses of the whole immune system to different antigens are not completely isolated from one another and that the history of encounters with different antigens plays a crucial role in the maturation of the whole system. Thus, problems of complexity, generation of diversity and self-organization have entered the field of immunology. Increasingly, an analogy has been drawn between the central nervous system and the immune system, both being viewed as networks of interacting cells producing cooperative behavior with the collective properties of memory and learning. It was realized that the immune system, like the central nervous system, could be studied as a computing network endowed with cognitive capabilities.

In the meantime, powerful computation methods have been developed in artificial intelligence and in the physics of complex systems; these have facilitated much improved simulations of such networks. (Some of them have already been presented in previous volumes of the Springer Series in Synergetics.) The above developments were the motivation for the symposium in Jerusalem in May, 1988, where physicists, computer scientists and mathematicians met with experimental immunologists. The goal was to review network approaches to the study of the immune system from different viewpoints corresponding to different levels of organization. Hence the title "Theories of Immune Networks".

The first part of the book presents empirical data on some molecular and cellular interactions, mainly idiotypic-antiidiotypic, which constitute the basis of the connections between different cell populations. Automata network behavior can be used to model not only idiotypic-antiidiotypic interactions, but also other specific and less specific interactions between different classes of lymphocytes as well as receptor-antigen interactions. Thus, the otherwise rather general concept of an immune network takes on a more precise meaning, depending on which level of description is used. Two different approaches are discussed in parts two and three respectively: The large network approach consisting of all possible idiotypic-antiidiotypic interactions with cross-reactivity between antigenic structures; and the small network approach, better fitted to represent and analyze specific data on limited interactions involved in the response to a single antigen as provided by experimental models.

Through the different contributions to the book, we hope that the reader will become familiar with the variety of formal approaches available today to analyze the behavior of such networks. Presented among them are the first applications of neural network computation to immunology.

Jerusalem
July 1989

H. Atlan
I.R. Cohen

Contents

Introduction to Immune Networks

H. Atlan[1] *and I.R. Cohen*[2]

[1]Department of Medical Biophysics, Hadassah University Hospital, Jerusalem 91120, Israel
[2]Department of Cell Biology, Weizmann Institute of Science, Rehovot 76100, Israel

From the very beginning the immune system has surprised immunologists with its unpredicted complexity. No one would underestimate the nervous system; its formidable hard-wired anatomy inspires awe, the experience of consciousness elicits wonder. The immune system, invented by nature to fight germs, seemed to hold no mystery. The idea of clonal selection satisfied expectations for a generation; the immune system needed for its regulation no greater sophistication than did the flushing toilet. The antigen merely pulled the chain, letting loose the production of specific antibodies; the antibodies themselves turned off their own production by neutralizing the antigen. Memory and the secondary response were akin to having more water in a particular tank waiting for a second pull of the chain by the specific antigen. The problem of self-tolerance was solved by outlawing self-recognition.

Since those naive times immunology has discovered helper T cells and (possibly) suppressor T cells, immune response genes and major histocompatibility complex molecules, differentiation in the thymus, interleukins, cytokines, Fc receptors, idiotypes and anti-idiotypes, and gene rearrangement and other somatic mechanisms of receptor diversity. Throughout all this unforetold gush of riches, orthodox Immunology has remained steadfast in the reductionist belief that solving the chemistry (structure as well as sequence) of a key set of molecules will be rewarded by illumination of the mystery of immune regulation. If we only knew how antigen is presented by MHC molecules, if we only knew the structure of the T cell receptor, if we only knew more about this molecule or that molecule, we would achieve enlightenment. Fortunately, we are getting to know more and more about these molecules, but unfortunately, we find ourselves no closer to enlightenment. Obviously no one would expect to understand the nervous system (memory, learning, behavior, consciousness, and so on) by knowing everything about cortical mapping, or the genetics, chemistry and physiology of neurotransmitters, or the biophysics of nerve conduction. It is clear to all that the higher order functions of the nervous system are generated by higher order interactions of the elements of the system. Awareness that such is probably the case for the immune system has motivated the convening of the present symposium.

Since Jerne's formulation that antigen receptors can play the role of antigens, the concept of an immune network has been used more as a verbal explanation of complex cooperative phenomena than as a source of real understanding. Indeed, idiotypic-anti-idiotypic interactions, similar to synapses between neurons, can form a molecular basis for specific connections between populations of lymphocytes. The network idea views the immune response as a cooperative interaction between different cell populations rather than as the isolated activation of a single clone. However, the mechanisms whereby the organization of these connections takes place and is responsible for integrated functions are poorly understood. Today, we seem to mimic the situation of the early stages of the cognitive sciences and artificial intelligence during the fifties and sixties. The pioneering work of McCulloch and Pitts had established the logical properties of idealized networks of neurons. Emergent properties and self-organization were studied from a general and abstract point of view but no real breakthrough could be achieved until significant progress was made empirically, at the local level of synaptic transmission and signal processing. The functional organization of the visual cortex, the actual anatomical details of connections between different areas in the brain are well known examples.

In our attempts at theorization, it is important at this stage to combine a large variety of strategies, top-down and bottom-up; on the one hand looking for deductive properties of general intuitions, such as that of a large and deep highly connected idiotypic-anti-idiotypic network; on the other, looking for more limited tools of bookkeeping and local interpretation of complex data on apparently paradoxical and counter-intuitive phenoma.

Of the first kind in this volume are the contributions of de Boer, Grossman, Weisbuch, and Segel and Perelson in part 2. Of the second kind are the articles of Kauffmann, Atlan and Hoffer-Snyder, Kurten, and Agur in part 3. The contributions of Cohen and Urbain et al. in part 1 confront us with examples of the intricacies in structures and functions encountered in real life. This natural complexity is, of course, what constitutes ultimately the real challenge for theorization.

One of the main topics recurring in part 2 is the irrelevance of the now classical picture of an immense and highly connected immune network. According to this representation, all the B and T cell populations of a repertoire would be interconnected by means of cross-reactivity on the one hand and "deep" idiotypic-anti-idiotypic-anti-anti-idiotypic-etc., connectivity on the other. Thus, the response of the immune system to any antigen would be that of the whole network. This picture is now under attack from different points of view. Without additional assumptions, percolation properties of such networks - i.e. the propagation of any stimulus throughout the whole system - do not allow for the establishment of the association of diversity and specificity actually observed in the immune response. On the other hand, significant modifications of the model are suggested by empirical data. Among others, one can mention the evidence of suppressor T cells, which may be more readily observed in the regulation of auto-immunity than in the response to foreign antigens and the reduction in connectivity of an initially highly connected B cell network during the maturation of the immune system. In the absence of suppressor cells, balance of growth by itself - i.e. the slowing down of cell proliferation during differentiation - can produce a suppressor effect. However, it is not clear yet to what extent this effect is a true inactivation or merely a decrease in stimulation, (i.e. an effect on the time derivative of proliferation rather than an actual suppression of activity). Anyway, both empirical data and theoretical analysis support the idea that specificity is likely to be achieved by subtle balances between activating and suppressing intercellular interactions.

Thus, the large and deep network may be seen at best as an initial and potential connection structure which subsequently organizes itself into functionally independent subsets. Such a process of self-organization is able to generate the high degree of specificity and diversity observed in real life, while being always dependent on the previous developmental history.

These results are reminiscent of work on neural networks carried out in the fifties, which demonstrated the necessity of inhibitory synapses to prevent the meaningless spread of any stimulus in the absence of strictly limited and predetermined specific connections [1]. Similarly, the crucial role of balances between excitatory and inhibitory connections was established in the production of non-trivial functional states [2].

For a long time, formal neural networks remained conceptual models with general logical properties rather than computation tools with practical applications. More recently, the tools developed jointly in the physics of disordered systems and in cognitive sciences have transformed neural networks into a new method of computation which extends the capacity of the usual kinetics by differential equations and kinetic logics. In particular, the notions of distributed memory and of learning are readily generalized to any kind of network of interconnected automata. The contributions of part 3 are preliminary attempts to apply this method to limited and circumscribed problems of immune regulation. Computational advantages and disadvantages of neural networks are compared with those of kinetic logics and continuous differential equations.

It is clear that much work has to be done to get a better insight into what can be expected from a generalization of this approach. However, one can be sure that immunological research at least as much as neuroscience, will benefit from the development of neural network computation facilities. In addition, contrary to the main trend in artificial intelligence, the analysis of small neural nets is relevant here. This may be a good motivation to explore more specifically the properties of such networks, small enough to be studied without having to resort to statistical methods. Their behavior can be described exhaustively and understood from the details of their structure. Yet, far from being trivial, their dynamics is rich enough to exhibit delocalized and cooperative properties characteristic of complex non linear systems.

Thus, from different approaches, a general picture seems to emerge; that of the immune system as an evolving network or self-organizing entity - a non programmed history of encounters with partially random internal and external antigenic stimuli is constantly integrated and serves to modulate and specify the more general initial genetic determinations. Among other classical problems, the mechanisms of self versus non-self discrimination are likely to benefit from a better, less anthropomorphic, understanding within this new paradigm.

Another result is a lesson of modesty. We are still in a situation where many different models have equivalent explanatory and predictive power; just because the complexity of the systems under study, even in the case of small regulatory subsets, is such that not enough empirical data are available - and it is doubtful that they will ever be - to restrict a vast body of possible models and to decide between equally plausible theories. In other words, as in the cognitive sciences, the underdetermination of the theories by the facts [3] prevents any belief in the ultimate truth of a fashionable theory. However, this remark is not meant to discourage research. On the contrary, one should be even more motivated for the necessary endless back and forth from modelling to experiment.

References

1. R.L. Beurle : Phil Trans. Roy. Soc. B., 240, 55 (1956)
2. W.S. McCulloch : Brookhaven Symposium in Biology 10, 207 (1957)
3. H. Atlan : Bull. Mathem. Biol., 51 (2) 247, (1989)

Part I

Natural Id-Anti-Id Networks and the Immunological Homunculus

I.R. Cohen

Department of Cell Biology, The Weizmann Institute of Science,
Rehovot 76100, Israel

It is postulated by Jerne and his associates that a network of mutually interacting idiotypes (ids) and anti-idiotypes (anti-ids) is a major factor in regulating the immune response [1]. Analogous to the nervous system wherein an environmental stimulus becomes information as a result of setting into motion self-connected neural networks, the environmental stimulus of the immune system, the antigen, acquires meaning by impinging on self-recognizing id-anti-id networks. Although the id-anti-id hypothesis does not explain or necessitate other demonstrably important immune system factors such as MHC molecules or helper T cells, the notion has fruitfully roused the neural networks of many immunologists.

The aim of this article is to state the lessons my colleagues and I have learned by studying id-anti-id networks of two sorts: one sort expressed by interacting antibodies, the other by interacting T cells. My aim is to draw attention to observations that might be considered by those proposing network solutions to immune regulation problems.

1. Id as Internal Signal or Id as Antigen

A fundamental distinction can be drawn between an id as a network connector or signal internal to the immune system, and an id as an antigen. An id may be perceived by the immune system as any other immunogenic macromolecular structure capable of stimulating the production of antibodies. To the extent that these antibodies are specific for the id, they may be termed anti-ids. In practice, the isolated id is often purified and injected in an aggregated form together with a strong adjuvant [2]. The resulting anti-id in turn is then isolated , purified and used with a suitable adjuvant to immunize yet a third set of animals, that in turn respond by making an anti-anti-id. There seems to be no end to the chain of antibodies that may be generated by such contrived immunization and this led Jerne to postulate in his original formulation that the idiotypic network was open-ended and proceeded until it fed back upon itself [1].

An id as network connector is quite another creature. Here one should be dealing with ids, anti-ids, anti-anti-ids, etc. arising in a single individual and accompanying or perhaps even preceding the response to a designated antigen. We have focused our studies on natural id networks. Natural networks, as we discovered, may differ considerably from contrived networks.

2. An Antibody Network: The Response to Insulin

My colleagues and I have investigated two types of natural id-anti-id networks related to insulin; that evoked spontaneously by immunizing mice or guinea pigs to ungulate insulins [3-7], and that appearing in mice, rats or humans spontaneously developing autoimmune diabetes. The details of these systems have been or are about to be published and I shall not describe the experiments here, only their meaning. Suffice it to say that the epitope that triggers the network is the portion of the insulin molecule bound by the insulin hormone receptor. This epitope is highly conserved, if not

Springer Series in Synergetics, Vol. 46 **Theories of Immune Networks**
Editor: H. Atlan and I.R. Cohen

identical in most mammalian species, with the notable exception of the guinea pig and her close relatives (the histricomorphs) who have a markedly aberrant insulin molecule [8].

The idiotype, which we have designated the DM-id in recognition of D. Elias and M. Rapaport who isolated the first DM positive monoclonal hybridomas, binds the conserved epitope and thus mimics the insulin hormone receptor [9] The specific anti-idiotype, the anti-DM-id, which arises spontaneously after the appearance of the DM-id, mimics the insulin epitope and binds to the insulin hormone receptor, and thus activates the biochemical effects of insulin itself. In vivo, the anti-DM-id causes hypoglycemia when it first appears, but after about a week it produces down-regulation and desensitization of insulin receptors. This causes peripheral resistance to insulin [10]. Thus, the DM network is significant pathologically as well as interesting immunologically. What has it taught us about immune regulation?

3. The Natural Network Is Closed

In contrast to the openness of contrived networks, the natural DM network seems to be closed; it does not appear to extend beyond the anti-id. It may be claimed that the anti-anti-id and the anti-anti-anti-id etc. occur but we miss them for lack of sufficiently sensitive probes. I can only answer that we looked for them in vain.

The natural DM network is closed in another sense; as far as we can tell it is inducible only once in the adulthood of otherwise healthy mice. The DM-id is produced only transiently, on days 6-13 of the primary response to immunization with insulin [4]. We never succeeded in detecting the DM-id subsequently despite hyperimmunization of the mice [11].

Similarly, the anti-DM-id was observed to appear only once, on days 24-40 after immunization [4]. It could not be induced a second time by repeated immunization of healthy mice to insulin [11].

Animals spontaneously developing diabetes such as NOD mice or BB rats spontaneously develop both the DM-id and the anti-DM-id [11]. Unlike healthy mice, these creatures persist in producing both DM-id and anti-DM-id. Thus, the development of autoimmune diabetes is associated with a persisting (dyregulated?) network.

4. The Immune Response is Partial to the DM-id

In contrast to contrived anti-ids, which may be induced by artificial immunization to apparently any id, upon immunization to insulin a spontanoous anti-id was detected only to the DM-id. Although insulin-immunized mice make a variety of different antibodies to insulin negative for the DM-id, no spontaneous anti-ids were observed for these DM-negative antibodies. The bias for the DM-id can not be explained by the fact that the DM-id is an autoantibody; most of the antibodies that mice make to ungulate insulins can be absorbed by mouse insulin. Thus, even the DM-negative anti-insulin antibodies include autoantibodies; nevertheless, DM-negative ids do not appear to be regulated by anti-ids.

5. The Natural DM Network is Conserved

The partiality of the immune system towards the DM-id is also evident in the cross-species conservation of this network. We detected the DM-id in mice of all strains responding to insulin, in rats [11], in guinea pigs [6,7] and in humans [12]. Indeed, mouse monoclonal DM-ids can be used as reagents for detecting human anti-DM-ids and mouse anti-DM-ids can be used to detect human DM-ids [9]. Thus the DM-id network is both dominant within a species and conserved in evolution.

6. The Immunological Homunculus

Why is this insulin antibody, the one we call the DM-id different from all other insulin antibodies? Implicit in this question is the realization that not all ids are created equal, at least in the eyes of the network. Since DM-positive and DM-negative ids may function equally as autoantibodies, that is they bind to self-insulin, we may also conclude that not all autoantibodies are dealt with by the same regulatory mechanisms. In the eyes of the immune system, why does one autoantibody enjoy special privileges?

In pondering this question it is worth noting the fact that guinea pigs upon immunization to ungulate insulins produce the DM-id; but unlike mice they don't turn off the DM-id or make an anti-DM-id [6,7]. Hyperimmunized guinea pig anti-insulin antiserum is rich in DM-id; anti-DM-id is undetectable. Recall that guinea pigs express an insulin molecule that has mutated away from the standard shape; it is studded with mutations in the conserved portion leading to a loss of about 99% of its ability to interact with the standard insulin hormone receptors characteristic of other mammals [8]. In short, guinea pig insulin lacks the DM epitope. As a consequence, guinea pig insulin cannot trigger the DM network in mice. Ungulate insulin does bear the DM epitope and so triggers the DM-id both in mice and in guinea pigs. But only mice go on to make the anti-DM-id. This suggests that it is not the mere presence of the DM-id that tells the network to make an anti-DM-id; rather the structure of the individual's own insulin (mouse versus guinea pig) bears the information. Thus, the network may be fashioned not around the immunizing epitope, but rather around the structure of the self; or to be more precise, around certain favored structures of the self.

To generalize beyond the response to insulin, natural anti-id networks seem to be more readily inducible by epitopes that are ligands for certain physiological receptors. For example, immunization to ligands of the acetylcholine receptor induces anti-ids that recognize the acetylcholine receptor [13]; immunization to thyroid stimulating hormone (TSH) induces anti-ids that recognize the TSH receptor [14]. In short, the natural network seems to favour structures composed of the active sites of physiological ligands so that the anti-id may behave as a ligand for a receptor (non-immune).

It is doubtful that there is a selective advantage in making an anti-id that acts like an anti-receptor antibody; on the contrary there is probably a physiological disadvantage. Therefore, it is conceivable that the network makes such an anti-id the better to control it, to guarantee that if it does get made, then it gets made only once. Recall that this is the case in normal mice. In contrast, diabetes prone mice and rats, and possibly humans, have a problem in regulating the anti-DM-id.

Be that as it may, the natural anti-DM-id is surely an internal immunological image of the functional portion of the individual's own insulin molecule. Other, non-DM domains of the individual's insulin are not represented in the network. Thus the network creates a highly selective representation of certain body structures. This recalls the homunculus, the "little man" engraved in the motor and sensory areas of the brain cortex. The picture of the little man is topologically distorted; it does not represent the space or volume occupied by the particular organ, but instead gives weight to the functional importance of the organ. (The homunculus in the human brain has giant thumbs and vocal cords, the dog's has a big nose). The brain uses its internal little man to sort and process nervous information.

The immune system also could, probably must have its little man to consult in processing information, particularly to aid in deciding what is self. As the insulins of humans and mice have a common DM epitope, they both share a common DM shape in the immunological homunculus. The guinea pig's immunological homunculus has a different picture of insulin because the guinea pig has a different self-insulin.

In addition to insulin, humans and mice share other macromolecular similarities, and this could explain idiotypic similarities between human and murine autoantibodies in systemic lupus erythematosis (SLE) or other autoimmune conditions. A new mouse model of SLE discovered by S. Mendlovic and his colleagues illustrates the importance of network interactions in the induction of autoimmunity [15]. Mice of a strain that does not spontaneously develop SLE were found to develop the disease along with its characteristic complement of various autoantibodies following immunization with a human monoclonal antibody bearing a common idiotype associated with human SLE. In other words, the anti-idiotypic response of the mice to the human-SLE idiotype unleashed SLE with all its immunologic manifestations. Is there a pre-formed SLE-network lying dormant in the immunological homunculus?

Obviously within each species there is even greater uniformity of functional macromolecules, and thus of the immunological homunculus. The common nature of the immunological homunculus may explain why humans with a common autoimmune disease produce autoantibodies to the same epitopes and, perhaps why these epitopes are often functionally important enzymes. For example, thyroid peroxidase is a major autoantigen in thyroiditis [16], lipoate acetyltransferase in primary biliary cirrhosis [17], and a cytochrome enzyme in chronic active hepatitis [18]. Anti-nuclear antibodies also may be directed against enzymes [19]. Is it an accident that the target antigens of autoimmune reactions are not only shared but functional?

7. The T Cell Network

To serve as a reference for interpretating and evaluating incoming antigenic signals, the immunological homunculus, like the neurological homunculus, must be formed before the antigenic signals enter the system. To put it another way; an epitope demonstrates its immunological dominance when it is preferred above alternative epitopes as the target for an immune response. The DM epitope is dominant, at least initially, because the immune system is receptive; the immunological homunculus anticipates the DM epitope.

How then is the immunological homunculus encoded? How can the DM-id know that it should be made even before immunization to insulin? To find an answer to this question we have begun to measure the responsiveness of T cells to monoclonal DM-id and anti-DM-id antibodies. More experiments are needed to draw firm conclusions, nevertheless the results thus far have raised the possibility that the immunological homunculus may be encoded in the reactivities of T cells.

The experiments involved culturing spleen or lymph node cells from mice with monoclonal DM-id or anti-DM-id antibodies and measuring the incorporation of labelled thymidine into DNA as a measure of T cell reactivity during various stages in the response to insulin. The results showed that naive mice had slight but significant T cell reactivity to the DM-id before being immunized to insulin. There was no reactivity to the anti-DM-id or to a DM-negative anti-insulin antibody. A week after immunization to insulin, at the time of the appearance of the DM-id antibody, there was an increase in T cell reactivity to the DM-id. This reactivity waned and disappeared as there arose T cell reactivity specific for the monoclonal anti-DM-id. This anti-(anti-DM-id) reactivity peaked at the time of the peak in anti-DM-id antibody and then it too declined; but it did not disappear. For at least 6 months after immunization to insulin, anti-(anti-DM-id) T cell reactivity was detectable. Anti-(DM-id) T cell reactivity was no longer observed during this time. Thus, the DM-id antibody was preceded by anti-(DM-id) T cell reactivity and the anti-DM-id antibody was accompanied by anti-(anti-DM-id) T cell reactivity. Persistence of the anti-(anti-DM-id) T cell reactivity was associated with resistance to reinduction of the DM network. In other words, the dominance of the DM network was associated with preexisting anti-(DM-id) T cells and permanent down-regulation of the DM network with persistence of anti-(anti-DM-id) T cells.

T cells, among their other functions, are regulators of antibody production by B cells. Therefore, it is reasonable to suppose that the behavior of the DM antibody network, perhaps even its existence, is founded on T cell activity. In a fundamental sense, the DM antibody network might be encoded in a T cell network.

My associates and I are presently studying a second example of immunological dominance associated with preexisting anti-idiotypic T cell reactivity, that related to the 65KD heat shock protein (hsp65). The hsp65 molecule is a major immunodominant antigen in Mycobacteria [20]. Persons immunized to M. tuberculosis or M. Leprae make immune responses primarily to epitopes on hsp65. This is unexpected because hsp65 is a highly conserved molecule and there is a very close sequence homology between mammalian and bacterial hsp65 molecules. Parts of bacterial hsp65 probably look like self-epitopes to the immune systems of mammals. Indeed, immunity to bacterial hsp65 is associated with autoimmune arthritis both in rats (adjuvant arthritis; [21]) and in humans (rheumatoid arthritis; [22]).

Mycobacteria express about 10^4 genes furnishing the mammalian immune system with a very large number of safely foreign epitopes. Why should the immune response focus with such consistency and vigor on hsp65, which looks like self to the extent that it may arouse autoimmunity? On the contrary, the principle of horror autotoxicus should lead one to expect hsp65 to be a very poor immunogen. (The fact that this expectation is contradicted by reality should by itself suggest that our basic ideas about self-tolerance may be in need of revision.)

To investigate the T cell immune response to hsp65 we developed a T cell line, designated M1, specific for the hsp65 molecule. M1 seems to exemplify a major shared anti-hsp65 T cell idiotype. Relevant to the present discussion is the observation that naive rats express a slight but significant degree of T cell reactivity, not to hsp65 itself, but to the anti-hsp65 M1 line. In other words anti-(anti-hsp65) T cell reactivity preexists, anticipates as it were, immunization to the mycobacterium and its hsp65 antigen. After immunization to the whole mycobacterium the response to the anti-hsp65 idiotype actually flares up sooner (by day 4) than does the response to the hsp65 molecule or other mycobacterial antigens. Later however (by day 10) the magnitude of the T cell response to the hsp65 antigen surpasses that of the T cell response to the M1 anti-hsp65 idiotype.

Thus, natural T cell anti-idiotypic reactivity precedes the immune response to the hsp65 antigen as it does the triggering of the DM antibody network. Hence, the observation of preexisting T cell anti-idiotypes is not unique to the response to the DM epitope of insulin and may, upon further investigation turn out to be a general phenomenon. Perhaps the immunological homunculus is encoded in such unsolicited T cell reactivity. H. Atlan and I have recently developed an automaton model of a T cell regulatory network integrating anti-idiotypic T cells, which recognize the effector T cell, with helper and suppressor T cells, which recognize the antigen [23]. A number of such regulatory units organized around key self-antigens could comprise the homunculus.

8. Problems and Paradoxes

The immunological homunculus described here is composed of a limited set of spontaneously reactive anti-idiotypic T cells that function to enhance the immunological dominance of certain self-epitopes or self-mimicking antigens. The result of this dominance is tight regulation, although dysregulation and autoimmune disease involving the dominant antigen occurs in some relatively few individuals. This formulation leaves us with a number of questions for experimental and theoretical consideration.

Deletion of autoreactive T cells is shown to take place, probably in the thymus [24]. How then can self-tolerance be regulated in practice by the apparently mutually exclusive mechanisms of deletion on the one hand and spontaneous heightened autoreactivity on the other hand? Are there fixed classes of self-antigens handled in one or the other way? If so, what decides which self-antigens are regulated by clonal deletion and which by anti-id networks?

It is not only the self-antigens that are problematic. T cells seem to recognize epitopes composed of relatively short peptide segments of the conventional antigen fixed in a cleft of an MHC molecule [25]. Does a regulatory anti-idiotypic T cell also recognize an MHC molecule with a small processed peptide segment of the T cell receptor or antibody idiotype, or does the T cell recognize the structure of the idiotype itself? If the latter were true, then the binding site of the anti-idiotypic T cell receptor would mimic the structure of antigenic epitope. Such a preexisting T cell could be understood to prime the response to the specific epitope. It would be more difficult to envision how a peptide-recognizing T cell could prime the immune system for a response to the antigen itself. As yet the anti-id T cells have not been cloned and we have no definitive answer to the id recognition question: peptide-MHC or unprocessed id. Nevertheless, the DM-id network does not appear to be MHC restricted - mice of all H-2 genotypes responding to Insulin make the same DM network; a finding which does not support, but also does not contradict the peptide-MHC view of T cell recognition.

9. T Cell Vaccination

Although much remains to be clarified regarding the functioning of anti-idiotypic T cells in regulating the immune system, it has been possible to mobilize such T cells to prevent or treat autoimmune disease. Autoimmune effector T cells responsible for mediating particular autoimmune diseases in experimental animals when suitably treated can be used as vaccines to elicit or augment the activity of specific anti-idiotypic T cells; a procedure termed T cell vaccination [26] . The vaccinated animals develop heightened anti-idiotypic T cell responses to the pathogenic T cell clones responsible for the disease and thereby the disease is prevented or suppressed [27]. The effectiveness of T cell vaccination in animal models has provided the rational for its application to human autoimmune disease [28]. One might reason that the usefulness of T cell vaccination for medicinal purposes rests on the natural physiology of T cell anti-idiotypy in immunological control.

References

1. N.K. Jerne: Ann. Immunol. (Inst. Pasteur) 125C, 373 (1974).
2. K. Sege, P.A. Peterson: Proc. Natl. Acad. Sci. USA 75, 2443 (1978).
3. Y. Schecter, R. Maron, D. Elias, I.R. Cohen: Science 216, 542 (1982).
4. Y. Schecter, D. Elias, R. Maron, I.R. Cohen: J. Biol. Chem. 259, 6411 (1984).
5. D. Elias, R. Maron, I.R. Cohen, Y. Shechter: J. Biol. Chem. 259, 6416 (1984).
6. I.R. Cohen, D. Elias, R. Maron, Y. Shechter: In Idiotypy in Biology and Medicine. ed. by H. Kohler, J. Urbain, P.A. Cazenave 20, 385 (Academic Press, New York, 1984).
7. Y. Shechter, D. Elias, R. Bruck, R. Maron, I.R. Cohen: In Anti-idiotypes, Receptors and Molecular Mimicry. ed. by D. S. Linthicum, N.R. Farid, 73 (Springer-Verlag, 1988).
8. R.W. Neville, B.J. Weir, N.R. Lazarus: Diabetes 22, 851 (1973).
9. I.R. Cohen, D. Elias, M. Rapopport, Y. Shechter: Methods in Enzymology (in press 1989).
10. D. Elias, M. Rapopport, I.R. Cohen, Y. Shechter: J. Clin. Invest. 81, 1979 (1988).
11. D. Elias et al in preparation.
12. D. Elias, I.R. Cohen, Y. Shechter, Z. Spirer, A. Golander: Diabetes 36, 348 (1987).
13. N.H. Wassermann et al: Proc. Natl. Acad. Sci. USA 79, 4810 (1982).
14. G.N. Beall, B. Rapoport, I.J. Chopra, S.R. Kruger: J. Clin. Invest. 75, 1435 (1985).
15. S. Mendlovic, S. Brocke, Y. Shoenfeld, M. Ben-Bassat, A. Meshorer, R. Bakimer, E. Mozes: Proc. Natl. Acad. Sci. USA. 85, 2260 (1988).
16. M. Lugate et al: J. Clin. Endocrinol. Metab. (in press).
17. S.J. Yeaman et al: Lancet I, 1067 (1988).
18. M. Guegen et al: J. Exp. Med. 168, 801 (1988).
19. J.H. Shero et al: Science 231, 737 (1987).
20. D. Young, R. Lathigra, R. Hendrix, D. Sweetser, R.A. Young: Proc. Natl. Acad. Sci. USA 85, 4267 (1988).
21. W. van Eden, J. Thole, R. Van Der Zee, A. Noordzij, J.D.A. van Embden, E.J. Hensen, I.R. Cohen: Nature 331, 171 (1988).
22. P.C.M. Res, C.G. Schaar, F.C. Breedveld, W. van Eden, J.D.A. van Embden, I.R. Cohen, R.R.P. de Vries: The Lancet II, 478 (1988).
23. I.R. Cohen, H. Atlan: J. Autoimmunity, in press.
24. H. von Boehmer, H.S. Teh, P. Kisielow: Immunology Today 10, 57 (1989).
25. P.J. Bjorkman et al: Nature 329, 506 (1987).
26. I,R. Cohen: Immunological Rev. 94, 5 (1986).
27. O. Lider, E. Beraud, T. Reshef, I.R. Cohen: Science 239, 181 (1988).
28. I.R. Cohen, H.L. Weiner: Immunology Today 9, 322 (1988).

Self-Nonself Immunological Tolerance and Idiotype Networks

J. Urbain, G. Urbain-Vansanten, and D. De Wit

Laboratoire de Physiologie Animale, Département de Biologie Moléculaire, Université Libre de Bruxelles, 67, rue des Chevaux, B-1640 Rhode Saint Genèse, Belgium

Introduction

The problem of self-nonself discrimination is one of the most fascinating in the immune system.

The mechanisms of the generation of diversity of immunological recognition units (Ig and TcR) endow a mouse with the ability of synthesising 10^9 antibodies or more. This is the potential repertoire. This potential repertoire is reduced by the fact that a single mouse contains less lymphocytes than the total number of possible antibodies. This will lead to the actual repertoire. The actual repertoire will be reduced by the silencing of the antiself repertoire.

If we suppose that even intracellular components are continuously exposed to the immune system (by processing and reappearance on the cell surface), the number of self antigens (the somatic self) should be around 10^5. This would lead, if one ignores the immune system itself (the immunological self) to something around 10^6 self epitopes. Since any immune response is degenerate, one could estimate that 10^7 clones should be silenced by the self-nonself discrimination.

All theories trying to explain tolerance have to rely on the principle that the only distinction between self antigen (S) and foreign antigens (F) is the continuous presence of S (they

Springer Series in Synergetics, Vol. 46 **Theories of Immune Networks**
Editor: H. Atlan and I.R. Cohen © Springer-Verlag Berlin, Heidelberg 1989

are there from the start and forever) while F antigens are present only in a transient fashion and in most cases are not there when the system starts. So tolerance must be due to a learning mechanism.

Actually two main groups of theories have been put forward : the temporal model and the spatial model.

In the temporal model, following the initial suggestions of Burnet and Ledeberg, immature lymphocytes are thought to pass through a differentiation stage in which they are especially sensitive to negative signalling if they meet an antigen. If they do not meet an antigen (which is the case for all B lymphocytes which are specific for foreign antigens), they will mature and join the pool of lymphocytes, waiting for antigen.

In fact, more than 10 years ago, several groups (Raff, Unanue, Urbain, Klinman) found that early B lymphocytes are especially sensitive to negative signalling.

This negative signalling could lead to clonal death or to clonal anergy (silencing). Clonal death corresponds to physical deletion and clonal anergy correponds to functional deletion.

The use of the word deletion alone is confusing and misleading. The processes of clonal death or anergy could be associated with the activation of a suppression maintenance mechanism which could inhibit those new B lymphocytes of the same specificity which appear continuously from stem cells within the bone marrow. Since the half life of most B lymphocytes is the order of a few days, there should be a maintenance mechanism which could cope with turnover problem.

In fact, in many papers, clonal deletion and suppression

mechanisms are viewed as antagonistic. As early as in 1975 we proposed that both processes should act in concert.
The second main framework to explain tolerance is the spatial model elaborated by Cohn and Bretscher. In this model, every B lymphocyte, independently of the stage of differentiation, is inducible or paralysable. More precisely, the interaction of a lymphocyte with the antigen only, leads to negative unresponsiveness (signal 1) while the meeting of antigen in conjunction with a second signal delivered by T helper cells leads to proliferation and differentiation of the B lymphocyte (signal 1 plus signal 2).
In this short review we examine the most recent data (including unpublished ones) to try to understand tolerance at both the T and B cell level.

Tolerance and T Lymphocytes

Using the most recent evidence available at the time of writing the following scheme of differentiation seems to occur for T lymphocytes. The most immature enter into subcortical area of the thymus . At that time they become CD 4^+ and CD 8^+. They have an immunological receptor, the heterodimer $\alpha\beta$.
If the receptor $\alpha\beta$ is able to interact with class I antigens (H2K, H2D, HLA-A, HLA-B, HLA-C...) the thymocyte is rescued from cell death (thymic death) and will become CD8 positive and CD4 negative. If the ab receptor has affinity for class II antigens, the thymocyte is rescued from cell death and will become CD4 positive and CD8 negative. This process is called the positive selection which takes place by interaction with the radioresistant thymic epithelium.

Although this process of thymic selection has been dismissed by some as being due to chimaera artefacts, the recent experiments using transgenic mice backcrossed to scid mice prove beyond doubt that this process does indeed occur as was originally claimed by Zinkernagel and others.

The second selection process eliminates autoreactive T cells and is supposed to occur when immature thymocytes meet self antigen carried by some cells of hematopoïetic origin. The details of this paradoxical double selection mechanisms are entirely unknown (see Janeway 1988).

It seems logical to assume that positive selection takes place before negative selection, otherwise we should have the problem of rescuing cells from nothing.

Let us now consider what we know about the elimination of autoreactive T cells. At first sight the results seems to be contradictory.

In one approach, Marrack, Kappler and their colleagues used the fact that many T lymphocytes, which use in their specific receptor the product of VB17 gene, do react whit IE of some haplotypes. Using an antibody which recognize specifically this kind of receptor, they look for the fate of T lymphocytes in various strains of mice or in chimaeras by FACS analysis. If the animal does express the particular IE, they could find lymphocytes displaying the VB17 product in immature double positive thymocytes ($CD4^+$, $CD8^+$) but not in single positive peripheral T lymphocytes ($CD4^+$ $CD8^-$;$CD8^+$ $CD4^-$). Administration of a monoclonal anti-IE antibody in the immature immune system blocks negative selection inasmuch as they can now find single positive peripheral T lymphocytes exhibiting the VB17 gene product. The autors conclude that autoreactive T cells are physically deleted. However the conclusion cannot be accepted as such since the absence of detection by FACS analysis cannot be equated with the physical absence of cells which could exhibit a reduction in the number of antiself receptors beyond the threshold of FACS sensitivity (see below).

The second approach which has been used by the group of von Boehmer makes use of transgenic mice. Briefly, they introduced into eggs a plasmid containing, besides an efficient promoter, one α gene and β gene obtained from a CTL clone recognizing the male antigen in the context of H2D of b haplotype. They then compared T cells from female or male transgenic mice carrying the b or other haplotypes. The male thymus was severely depleted and they could not find by FACS analysis antiself double positive thymocytes. In the periphery, they found high numbers of T lymphocytes which were bearing the transgene products (antiself) but which were inactive because the expression of the accessory molecule CD8 was severely reduced. The authors conclude that tolerance is due to a physical deletion of autoreactive clones in the male. The peripheral T cells which express the transgene products could have escaped the deletion process because the expression of the CD8 was reduced or absent. However one could suggest another possibility, namely that tolerance is not due to clonal death but to clonal anergy. The silencing mechanisms could be the inhibition of the synthesis of the necessary CD8 molecule. Of course, the transgenic model is far remote from normal physiology because the cells harbor a large number of transgenic copies placed under the control of a powerful promoter.

The contradictions seen at the level of the double positive positive between the two approaches could be explained by kinetic problems.

To summarize, the authors of the experiments conclude that clonal death is the principal cause of tolerance but, as we suggested, there are several reasons that might modify these conclusions.

Tolerance at the B Cell Level

Pioneering studies by several groups including our own, indicated that early B lymphocytes (from foetus or neonates) had indeed a different physiology from adult B lymphocytes. It was shown that early B lymphocytes, exposed to anti-μ antibodies or antigen seem to be unable to resynthesize their receptors after capping and endocytosis. At that time, it was impossible to evaluate the fate of "tolerant lymphocytes" : clonal death or survival but decreased density of Ig receptors. Later it was shown that this sensitivity to negative signalling could be avoided by inhibitors of protein synthesis like puromycin. So, the synthesis of something was needed to achieve negative signalling. Instead of describing the heroic efforts made over more than 10 years to understand this something, we shall describe two recent approaches to unravel the enigma of tolerance.

One approach (Goodnow et al. 1988) use, once again transgenic mice. A series of egg mice were injected with a plasmid

containing the gene for hen lyzozyme. Transgenic mice were obtained in which hen lyzozyme was now a self antigen, being continuously present from the start of the immune system. Indeed, the adult mice are tolerant to hen lyzozyme inasmuch as they do not respond to lyzozyme. Of particular interest is the fact that tolerance cannot be simply explained by the silencing of T helper cells since B lymphocytes from tolerant transgenic mice transferred into irradiated non transgenic mice and mixed with anti-lyzozyme T helper cells do not respond (see below).
A second set of transgenic mice was constructed. Eggs were injected with a plasmid containing the genes for one high affinity antilyzozyme antibody (the genes of the heavy chain contain V, D, J segments and a Cl- Cd region to allow their expression in immature B lymphocytes).
These transgenic mice synthesize constitutively antilyzozyme antibody (no lyzozyme injection is necessary). The two sets of transgenic mice were then crossed. So, potentially the F1 can make simultaneously a foreign antigen now treated as self (lyzozyme) and an antibody, which is now an antiself antibody. In the F1 animals, the synthesis of antilyzozyme antibodies is now suppressed. The mice are tolerant, as expected from previous knowledge. If one looks for B lymphocytes in tolerant mice by fluorescence, large number of B lymphocytes displaying antilyzozyme receptors are seen. These "silent" B lymphocytes display an abnormal phenotype. The surface IgM is greatly reduced and the level of IgD is high. In this kind of experiments, the extent of clonal death cannot be estimated but it is clear that most of the B lymphocytes displaying a high

affinity antiself receptor (10^9 L/M) have not been physically deleted. They are present and silent.

A second approach has been used by our group. We tried to achieve tolerance in a well known idiotypic system, the arsonate system. Briefly gravid mice are injected with arsonate coupled to mouse immunoglobulins or human Ig.

As a result, the neonates are exposed to oral induction of tolerance (by the milk). A long lasting tolerance could be achieved with arsonate coupled to human Ig, but not to mouse Ig!

Then the spleens of tolerant and control mice were subjected to a limiting dilution analysis after a polyclonal activation with LPS.

This procedure allows one to estimate the frequency of B lymphocytes potentially able to synthesize antiarsonate antibodies. The results suggest that tolerant mice do not exhibit a massive clonal death since the frequency of B lymphocytes able to make antiarsonate antibodies in tolerant mice is only twofold lower than in controls. Furthermore, tolerant spleens transferred into irradiated syngeneic recipients and challenged with antigen synthesize large amounts of antiarsonate antibodies.

Other experiments that we cannot describe here indicate that tolerance can be achieved in B lymphocytes even if helper T cells are functionally deleted.

Therefore, it is clear, as was already stressed by Nossal, that clonal anergy does indeed exist.

Tolerant lymphocytes do not necessarily die. They can survive and have stored "negative unresponsiveness". The deletion that we observe in limiting dilution analysis can be explained in two ways :

a) those B lymphocytes displaying the highest affinity receptors have really been clonally deleted.

b) those "anergic" B lymphocytes cannot be rescued by a polyconal activator like LPS.

By and large, if we compare the two approaches (transgenic mice and tolerant mice by perinatal exposure to antigen), the results do agree remarkably. It can also be seen that the results on B lymphocytes are not dramatically different from those of the group of von Boehmer on antiself T lymphocytes in transgenic mice.

Conclusion, or Rather, Comments on the Present Status

Many papers give the impression that tolerance should be explained in terms of clonal death or clonal anergy or suppression.

The problem is that "or" is a confusing word (it can be inclusive or exclusive). It is obvious that a complex and necessary biological system should rely on a hierarchy of mechanisms. It could well be that the antiself clones could be silenced (change of phenotype, decrease of receptor density, absence of accessory molecules ...).

Suppressive mechanisms could exist to maintain and control the state of tolerance. If, as suggested above, tolerance at the B cell level can be due to both clonal death and clonal anergy, we must also explain how the bone marrow precursors of the same specificity which emerge all the time (normal B lymphocytes have a half life of a few days) are silenced.

From the data of von Boehmer and colleagues, it appears that tolerance can exist and be maintained in the absence of suppressor clones. They can observe tolerance to self (HY) in "monoclonal mice" (transgenic mice backrossed to scid mice) possessing only a single clone of antiself lymphocytes. However one cannot conclude that physiological tolerance does not rely partly on suppressive mechanisms.

The Final Paradox

At the beginning of this paper, we tried to estimate the size of the antisomatic self repertoire, excluding immunological receptors (the immunological self).

We estimated that the somatic self could silence 10^7 clones. If we now suppose that the immunological self is treated in the same way as the somatic self, we get into serious troubles, since the immunological self contains 10^9 to 10^{10} receptors. The immune system cannot avoid the recognition of itself and therefore, given the number of self epitopes, the whole immune system should be silenced as soon as it is born.

We have analyzed this problem in details in a recent paper and have proposed "The Broken Mirror hypothesis". We shall only sketch the concept briefly here.

We assume that early B lymphocytes are not silenced by the simple encountering of B lymphocytes with a self antigen but rather upon meeting a self antigen sitting on the membrane of an early APC, the SPC (Self Presenting Cell).

At that time, there is no endogeneous synthesis of idiotypes. Therefore the immune system is born, non tolerant to most idiotypes. Only maternal idiotypes will be treated as self. We assume than that silenced B lymphocytes have a change in their phenotype. They will not transform into normal B lymphocytes but rather will become regulatory long lived B lymphocytes, unable to transform into antibody producing cells. A possible role for these regulatory B lymphocytes could be to select positively those normal B lymphocytes displaying complementary receptors or antiantiself.

Therefore a first stable repertoire is established and this first repertoire is shaped by the previsible environment (the self). This first repertoire will shape a second repertoire, the antiantiself which will become the available antiforeign repertoire.

BIBLIOGRAPHY.

Goodnow C.C., Crosbie J., Adelstein S. et al. (1988) - Nature 334,676.

Kappler J.W;, Roehm N., Marrack P. (1987) - Cell 49,273.

Marrack P., Kappler J., (1988) - Immunology Today 9,308.

Nossal G.J.V., (1983) - Ann. Rev. Immunology 1,33

Von Boehmer H., TEH H.S. and Kisielow P. (1989) - Immunology Today 10,57.

Urbain J. et al. (1988) - Ann. Inst. Pasteur 139,609.

Part II

Extensive Percolation in Reasonable Idiotypic Networks

R.J. De Boer

Bioinformatics Group, University of Utrecht, Padualaan 8,
NL-3584 CH Utrecht, The Netherlands

1. Abstract

We analyse a mathematical model specifying profound networks of idiotypic interactions amongst B cell clones. In the model we incorporate only fundamental assumptions about idiotypic interactions. Nevertheless, this "reasonable" idiotypic network model displays very "unreasonable" behaviour. As soon as the network is formulated as a profound network structure it displays "extensive percolation" of idiotypic signals. Thus the first antigenic perturbation of the network is expected to affect most of the clones. We have attempted to solve these problems by extending the model with seemingly important theoretical and/or immunological insights, but as yet we have failed. Therefore, our present interpretation of these "unrealistic" results is that immune systems do not function by means of profound network structures.

2. Introduction

Jerne's [1] idea that immune systems are regulated by networks of idiotypic interactions has roused considerable interest among both theoreticians and experimentalists. Immune systems perform complex functions such as specific immune reactions, the specific memory phenomenon, and self non-self discrimination. Although it remains an open question, it is conceivable that immune systems perform these tasks by means of these proposed network properties. To this end, immune systems have often been compared to neural networks [1-3]. Neural networks, for which Hopfield equations provide the general paradigm system, are capable of complex computational processes (such as learning and memory) [4]. The similarity between neural and idiotypic networks, and the powerful computational properties of neural networks, have probably shaped our intuition: in immunology theoreticians are generally inclined to think that idiotypic interactions do indeed play an important role in immunoregulation.

Because immunoregulation by means of these network properties is little more than an assumption, we here analyse it by means of a "bottom up" approach. Starting with basic assumptions about idiotypic interactions, and their effects on B cell proliferation, we develop a "fundamental" idiotypic network model. Such a model might eventually become comparable to neural network models, i.e. might eventually obtain comparable (immuno)computational properties. We thus investigate whether simple idiotypic interactions amongst B cell clones can indeed account for the proposed regulatory network properties. Our previous analysis [5-7] gave an ambiguous answer to this

Springer Series in Synergetics, Vol. 46 **Theories of Immune Networks**
Editor: H. Atlan and I.R. Cohen

question. On the one hand, it was shown that our idiotypic networks could not account for the control of proliferation. Proliferating clones in the network suppress their idiotypic partners long before these become sufficiently suppressive to control the proliferation. On the other hand, it was demonstrated that idiotypic interactions can easily account for memory phenomena: the interaction between an idiotypic and an anti-idiotypic clone brings about three different equilibria, one of which corresponds to an immune state for the idiotypic clone. Thus idiotypic network theory accounted for immunological memory, but could not account for proliferation regulation (i.e. for suppression) [6]. We will here test the robustness of these memory phenomena by analysing profound networks consisting of many (i.e. 100) instead of just a few (i.e. 2) clones.

3. The assumptions

We here consider clones of B-lymphocytes. We assume that the B-lymphocyte populations are regulated by three processes: 1) influx of newborn cells from the bone marrow, 2) normal turnover of cells, and 3) proliferation. It is further assumed that idiotypic interactions influence the rate of cell proliferation. The network consists of N clones, each of which is identified by a unique random receptor. A clone (X_i) recognises another clone (X_j) if (part of) their respective receptors can be matched complementarily. The accuracy of this match specifies the affinity (A_{ij}) of the idiotypic interaction between X_i and X_j ($0 \leq A_{ij} \leq 1$). Each clone can also be stimulated by its antigen (Ag_i).

Influx/efflux. In the absence of idiotypic interactions the clone size is determined solely by the balance between the source (S_i) of cells from bone marrow and the death (D) of cells in the periphery. This suffices for a stable virgin state at a clone size of S/D. Because the influx of cells per clone is small, and the rate of cell turnover is high, virgin populations are typically small. In the virgin state idiotypic interactions are therefore weak or absent.

Symmetry. If idiotypic recognition is based on complementary matching and receptor crosslinking, idiotypic interactions are necessarily symmetric. If idiotype "i" matches "j", "j" should also match "i"; if "i" antibodies crosslink "j" receptors, "j" antibodies can do the same. Hoffmann [8,9] first proposed this simple and attractive symmetry theory; he [9] and Jerne [10] review empirical data that support symmetry theory (Jerne's "preferred partners"). Clones never recognise themselves (i.e. we set all A_{ii} to zero).

Dose response relation. Because B cells are activated by the crosslinking of the antigen receptors, it is to be expected that the rate of cell activation will increase if the concentration of the crosslinking agent (here antigen or anti-idiotypic antibody) increases. However, when these concentrations become too high the efficacy of cell activation by receptor crosslinking decreases [11]. Thus, the present model is based on the efficacy of cell activation and proliferation induction (by receptor crosslinking); inhibitory interactions are incorporated as the reduction of the proliferation rate. This argument corresponds to a bell-shaped proliferation dose response curve, i.e. the kind that can be found in any immunology textbook.

The model. The rate of cell proliferation is governed by a growth dose response function $G(X_i, Ag_i, \alpha Id_i)$, which depends on 1) the size of the clone (a buffering term), 2) the antigen, and 3) the total amount of anti-idiotype (αId_i). The function $G(X_i, Ag_i, \alpha Id_i)$ is maximally one; proliferation per cell then proceeds at a rate P per day. The function (eq. 2) is a log bell-shaped curve: anti-idiotype concentrations can become too high. This reduces the crosslinking and hence the rate of cell proliferation. For reasons of simplicity, antigen can only increase (up to a certain maximum) the rate of cell proliferation: the antigen dose response curve is a simple saturation function. Antigen (Ag_i) cannot grow and is either removed by the clone that recognises it (X_i), or is incorporated as a constant (K=0). We thus propose the following model:

$$\alpha Id_i = \sum_{j=1}^{N} A_{ij} \circ X_j \qquad (1)$$

$$G(X_i, Ag_i, \alpha Id_i) = \frac{Ag_i + \alpha Id_i}{P_1 + F \circ X_i + Ag_i + \alpha Id_i} \circ \frac{P_2}{P_2 + \alpha Id_i} \qquad (2)$$

$$X_i' = S_i - D \circ X_i + P \circ X_i \circ G(X_i, Ag_i, \alpha Id_i) \qquad (3)$$

$$Ag_i' = -\frac{K \circ Ag_i \circ X_i}{K_1 + X_i} \qquad (4)$$

Parameters. The parameter setting is: $S \approx 10$ cells d^{-1}, $D=1$ d^{-1}, $P=1.5$ d^{-1}, $P_1=10^3$, $P_2=10^6$, $F=0.01$, $K=1$, $K_1=10^5$. Buffering ensures that large X_i populations ($F \circ X_i \approx P_1$) cannot be stimulated by small antigen and/or anti-idiotype concentrations [6]. The virgin population density equals $S/D \approx 10$ cells. The influx is slightly different for each clone (to prevent settlement into unstable equilibria): S has a mean of 10 cells per day with a 10% standard deviation. Virgin populations are too small to evoke proliferation ($S/D << P_1$): idiotypic interactions are negligible in the virgin state. Maximum proliferation proceeds at a rate of P-D=0.5 cells per cell per day (this corresponds to a doubling time of about 16 hours).

Methods. The affinity matrix is either based on a complementary matching algorithm, or is drawn randomly from a uniform distribution between zero and one. The complementary matching algorithm resembles that of Farmer *et al.* [12] and Perelson [13]. Ours matches patterns (landscapes) defined by an array of real numbers (that are uniformly distributed between zero and one). The algorithm is described elsewhere [7]. The low-D models have been analysed by GRIND [14] which performs numerical 0-isocline analysis and numerical integration by means of ROW4A [15]. High-D models are integrated by means of a variable step size Runge-Kutta-Merson integrator implemented in NAG [16].

4. Results and Discussion

Memory. We now summarise our previous analysis of immunological memory [6]. Consider a network of two clones: X_1 and X_2 that see each other with maximum affinity (i.e. $A_{12}=1$). This network will be stimulated with the antigen that is recognised by X_1 (i.e. Ag_1). In Fig.1a this 2-D network is analysed statically: the curved lines are the $X_1'=0$ and the $X_2'=0$ isoclines. The region in which X_1 expands is shaded; X_1 proliferates in the large shaded region situated at an intermediate X_2 population size. The isoclines intersect in three stable equilibria: the virgin state (V) and two immune states (I_1 and I_2). In the absence of antigen the system always remains in the virgin state; here idiotypic interactions are negligible. Fig. 1b shows the same picture with a trajectory (i.e. the fat line) that was initiated by the introduction of antigen ($Ag_1=10^4$). In response to antigen, X_1 proliferates (the trajectory moves to the right) which, in turn, evokes the proliferation of X_2 (the trajectory moves upwards). As a consequence of the proliferation of X_1, antigen is rejected (not shown) and the system settles into the I_1 equilibrium. In this new state the system is immune to Ag_1: reintroduction of Ag_1 leads to rapid rejection because X_1 is already enlarged. Conversely, X_2 is suppressed in the I_1 state: introduction of Ag_2 into this state never leads to rejection because X_2 fails to proliferate. Thus the previous exposure to Ag_1 is specifically remembered by the network.

In both immune states both clones are enlarged, i.e. both clones proliferate. Thus the network maintains its immunity to Ag_1 by the mutual stimulation between X_1 and X_2. The reason why X_2 is nevertheless said to be "suppressed" in the I_1 state is that it cannot proliferate any further. Any increase in X_1 decreases the rate of X_2 proliferation, whereas increasing X_2 increases X_1 proliferation. Antigenic stimulation of X_2 also fails to evoke additional proliferation and antigen cannot be rejected: X_2 is suppressed.

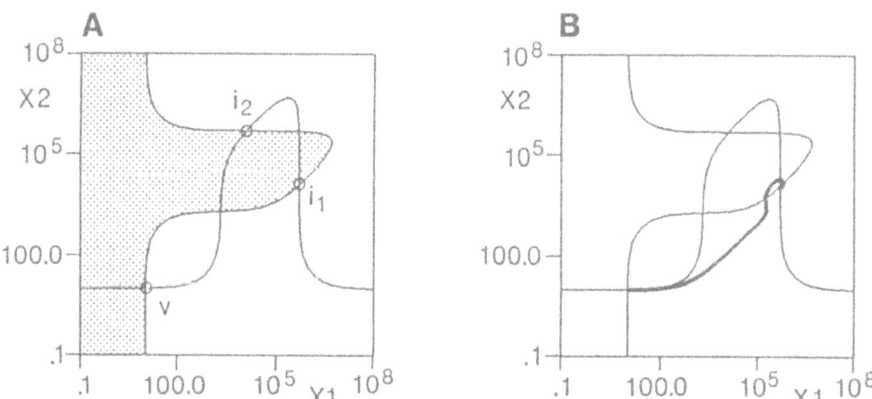

Figure 1. A 2-D network with maximum affinity ($A_{12}=1$). Fig. 2a: the $X_1'=0$ and $X_2'=0$ isoclines define 3 stable equilibria: a virgin state (V) and two immune states (I_1 and I_2). The $X_1'>0$ region is shaded. Fig. 2b: the trajectory of a switch from the virgin state to an immune state (I_1) as it is evoked by an antigen dose of $Ag_1=10^4$ cells.

Absence of fading. Now consider a system with a third clone (X_3) that interacts with X_2. We again introduce Ag_1 and, as a consequence, X_1 and X_2 switch to the I_1 equilibrium. In the I_1 equilibrium the anti-idiotypic X_2 population maintains the proliferation of the large idiotypic X_1 population (note that $X_1 >> X_2$). It is therefore to be expected that X_2 should also be able to initiate the proliferation of the (small, i.e. virgin) X_3 population. In our model we have implicitly (and quite reasonably) assumed that it should be easier to activate all cells of a small clone than of a large clone. Thus, on reasonable grounds, we expect X_2 to initiate the proliferation of X_3. X_3 as a consequence switches to an immune state that is comparable to that of X_1 in the I_1 state (not shown). This argument can however be continued: X_3 is expected to initiate the proliferation of the clone(s) to which it is connected, and so on. We conclude that the idiotypic activation signal fails to fade during its propagation into the network: along the propagation pathways clones keep on switching to "immune" or "suppressed" states comparable to those described above. Elsewhere [7] we provide more evidence for this argument.

Connectivity thresholds. The percolation of the idiotypic signal through the network depends not only on its extinction rate (i.e. on the above-mentioned fading), but also on the connectivity of the network. In highly connected networks the propagation pathway will branch several times, and the signal may fade away on one branch but proceed via another. We can analyse the connectivity properties of an idiotypic network by means of graph theory. A randomly connected, symmetric, idiotypic network corresponds to a (isotrophic) random undirected graph in which the idiotypic interactions are the edges (E) and in which the clones correspond to nodes (N); see also Perelson [13]. The connectivity properties of isotrophic random undirected graphs (in which E edges connect N nodes equiprobably) were analysed by Erdos & Renyi [17,18], in Kauffman [19]. Several results of random infinite graphs [17-19] are of interest for our (finite) random idiotypic networks. If E<<N only small isolated structures are found, in which any node is connected to a few others. Whenever the ratio of E/N exceeds 1/2, a threshold is reached and most nodes are interconnected in one enormous structure. As E/N increases still further, more isolated nodes are incorporated into this very large connected structure. (For finite graphs these thresholds soften to sigmoids). Note that in our idiotypic networks, an average of one edge per two clones (E/N=1/2) corresponds to an average of one idiotypic connection per clone.

The Erdos & Renyi theory thus predicts that once the connectivity exceeds one idiotypic connection per clone, most clones suddenly become interconnected. Such connectivity however is extremely weak: if the connectivity of the idiotypic network were to be still weaker, a number of clones would not be connected to the network at all. They would hence require other immunoregulatory mechanisms. Moreover empirical data suggest that one clone can be connected to very many (e.g. 40) clones [20-22]. It thus seems safe to assume that the network connectivity usually exceeds one connection per clone. According to the Erdos & Renyi theory such an assumption however means that most clones are interconnected in one large structure. Moreover, if we combine this conclusiom with our "absence of fading" results, it follows that each perturbation of the network (by e.g. antigen) eventually affects all the clones. This is very unreasonable: any antigenic stimulation of the immune system eventually forces almost all clones of the idiotypic network to switch to the "immune" or "suppressed" state. This unreasonable result, i.e. extensive percolation, however seems to be based on very reasonable assumptions and theories. At present we can see three possible explanations for this: 1) the Erdos & Renyi theory is not applicable to biotic idiotypic network structures, 2) we have omitted an essential element of the idiotypic network from our models, and/or 3) immune systems do not function as an idiotypic network.

1. Random vs. Pattern networks. The Erdos & Renyi theory is based upon random graphs. The connections in an idiotypic network are however not necessarily random. We here investigate the impact of the presupposed randomness on the percolation results. We compare networks based on random connections of a random strength ($0 \leq A_{ij} \leq 1$) with networks based on our complementary matching algorithm. The algorithm matches idiotypes that were drawn randomly; the resulting affinities are also scaled between 0 and 1. We refer to them as the "random" and the "pattern" networks respectively. In both networks the connectivity is measured by the average number of interactions per clone (nc). In "pattern" networks the connectivity is increased by increasing the size of the pattern (i.e. by allowing for more idiotopes per idiotype). We thus simulate networks of 100 clones; we perturb the network with antigen at day 100 (i.e. when it has attained the virgin state); antigen is removed after 25 days.

Two typical examples (of "pattern" networks) are shown in Fig. 2: one is located around the Erdos & Reyni threshold (nc=1.04, Fig. 2a) and the other slightly above it (nc=1.36, Fig. 2b). This slight difference in connectivity accounts for an enormous difference in behaviour: in Fig. 2a the immune reaction involves 3 clones whereas that of Fig. 2b involves 34 clones. This is mainly due to the

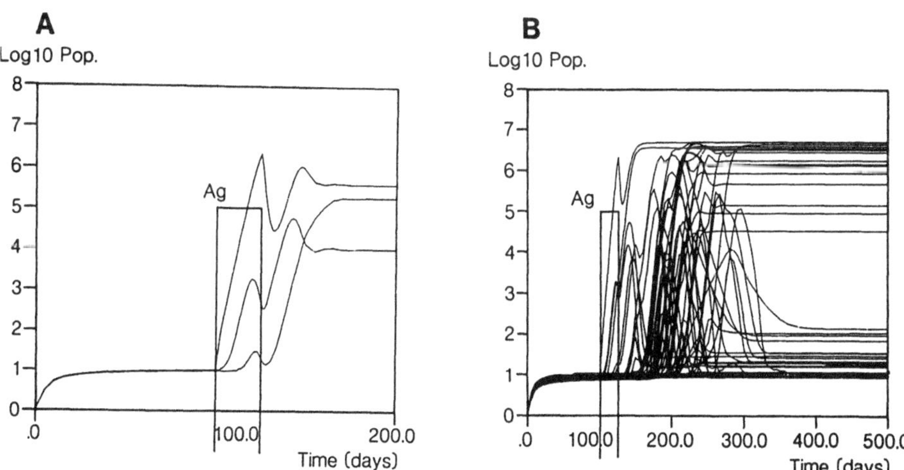

Figure 2. Examples of the behaviour of 100-D "pattern" networks. A network with on average 1.04 connections per clone (2a), and (2b) a network with 1.36 connections per clone. This is just around and just above the Erdos & Reyni threshold. The Figure only depicts the clones that are actually connected to the network stimulated by antigen (all other clones remain virgin).

connectivity properties: the clone that responded to the antigen was actually connected to 3 clones in Fig. 2a and to 55 clones in Fig. 2b. In Fig. 3 we score similar network properties as a function of the connectivity of the network. We score 1) the "effective" dimension of the network, i.e. the number of clones that are actually connected to the clone stimulated by antigen. And, in the equilibrium that is finally reached, we score 2) the number of clones that are actually affected, i.e. that have left the virgin state, and 3) the depth of signal propagation, i.e. the maximum idiotypic level affected. The first score measures the Erdos & Reyni property; the other two link this with "absence of fading".

The effective dimension curve (Fig. 3a,b: ❑) demonstrates the connectivity to switch from small isolated structures to one large interconnected network around one connection per clone. The threshold is relatively smooth, but for the almost infinite (10^7 clones) biotic immune system it is expected to be very steep. The number of clones that are actually connected to the clone stimulated by antigen (i.e. the effective dimension) hardly differ for "random" and "pattern" networks (Fig. 3a vs. 3b). We conclude that the Erdos & Reyni results are applicable to our idiotypic networks which are based on the very reasonable complementary matching assumption.

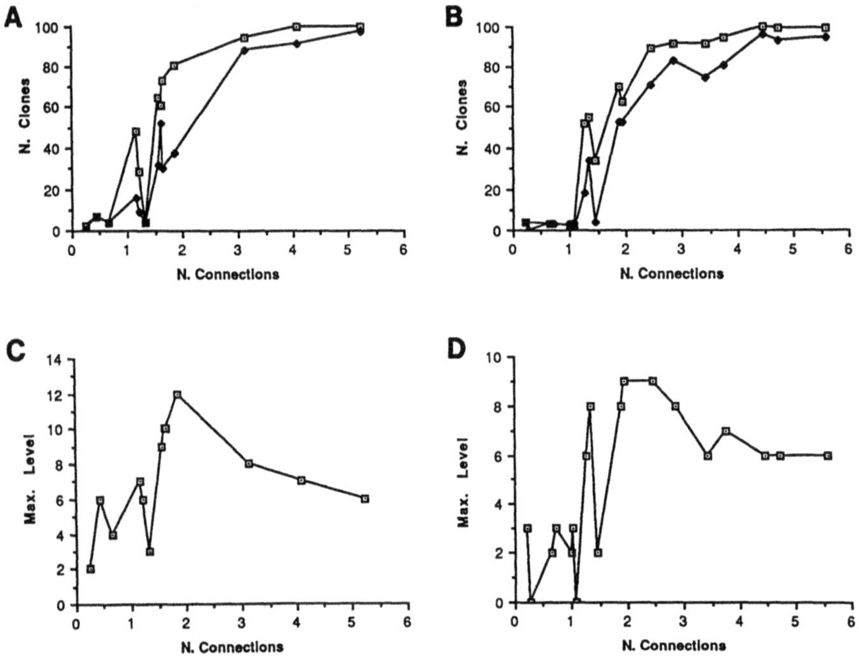

Figure 3. The network properties of 100-D "random" (3a,c) and "pattern" (3b,d) networks as a function of the average number of connections per clone. Fig. 3a,b: the actual number of clones connected to the clone stimulated by antigen (❑), and (◊) the actual number of clones that become affected by the signal, i.e. that leave the virgin state. Fig. 3c,d: the maximum idiotypic level reached by the idiotypic signal.

The two other properties that we score are also very similar for "random" and "pattern" networks. The number of affected clones (Fig. 3a,b: ◊) simply follows the effective dimension curve: a limited number of clones that are actually connected to the perturbed sub-network remain in the virgin state. Networks in which most clones have been affected by a previous perturbation are highly unresponsive to novel perturbations (i.e. to other antigens): antigens are either directly rejected because "their" clone is already immune or cannot be rejected because "their" clone is suppressed. Moreover, the percentage of suppressed clones increases if the network connectivity increases [5]. The maximum idiotypic level that is actually reached by the signal has a peak around 2-3 connections per clone. The maximum is about 10-12, but in 200-D systems we found higher values (e.g. 17). It takes the idiotypic signal several months to attain such a high idiotypic level (see Fig. 2b). During these months the network oscillates heavily, i.e. seems chaotic; we think such extensive oscillations are unrealistic.

2. What is missing? The fact that we find unrealistic results with simple networks that incorporate only "realistic" assumptions might suggest that "apparently" we have omitted an essential element of the idiotypic network. We here investigate this possibility. Note that if we indeed were to find a process that solved our percolation problem this would pinpoint this process as playing an essential role in the control of percolation. This would be a very important result since, so far, idiotypic network theory apparently lacks such a regulatory control structure.

Long-range inhibition. Segel & Perelson [23] and Segel [this volume] develop an idiotypic network model that incorporates "long-range inhibition" and "short-range activation"; this combination is known to promote pattern formation. In our model, activation and inhibition are governed by the same affinity matrix, i.e. both interactions are of an identical range. We have, however, tested the "long-range inhibition" idea in our "pattern" networks [7]. We made suppressive interactions more general by requiring a shorter complementary match (i.e. a smaller idiotope) for suppression than for activation. This can even be defended on immunological grounds: it is conceivable that anti-idiotypic antibodies that fail to crosslink idiotypic receptors might have sufficient affinity to these receptors to reduce (e.g. by temporary binding) the crosslinking process of another (high affinity) antibody. However, in our symmetric networks "long-range inhibition" fails to reduce the percolation. We find that the networks 1) can only slowly settle into equilibria, 2) seem more chaotic, and 3) still display extensive percolation. These counter-intuitive results are probably due to the fact that our networks do not discriminate between "activators" and "inhibitors". Whether a clone activates or inhibits depends on the idiotypic circumstances. Thus "long-range inhibition" can, after a long period of rather constant circumstances, inhibit a clone from inhibiting. As a consequence of these changing circumstances that clone not only stops inhibiting but may even start to activate its idiotypic partners. Thus "long-range inhibition" seems to imply "long-range activation".

Antibodies. In our simple model we have lumped into one population (X) B-lymphocytes and the antibodies that these cells produce. The elimination of antigen and the idiotypic communication between clones (i.e. the transfer of the idiotypic signal) in fact occur via antibody molecules. Additionally, cells only differentiate into antibody producing plasma cells after they have proliferated [24]; this accounts for a delay in the idiotypic communication. Moreover, antibodies have a longer life span than cells [10]. Elsewhere [7], we have incorporated these additional features in our idiotypic network model. Instead of solving our percolation problems, the addition of these features generated new unrealistic results. The network no longer switches from virgin to immune states (that still exist), but, following the introduction and elimination of antigen, oscillates heavily from the I_1 to the I_2 state, and vice versa. This cycle never ends. It is sustained by the slow decay of the antibodies: high, i.e. suppressive, concentrations of antibody decay slowly because they have suppressed the network. This decrease in the antibody concentration however changes the idiotypic circumstances, i.e. the antibody concentration becomes stimulatory again. Thus cells initiate proliferation and start to produce new antibodies until the antibody concentrations again become suppressive. We have also incorporated the formation of complexes of idiotypic and anti-idiotypic antibody molecules. This failed to solve the percolation problem; moreover the above-mentioned cycle persisted.

T cells. Idiotypic network theory was originally presented in terms of receptor interactions between B cells [1]; our model also considers B cells. T cells however play a crucial role in immunoregulation. Idiotypic network theory can be extended with T cells in two ways. First, B cell proliferation and antibody production usually result from B-T cell cooperation. B cells present antigen (and/or idiotype) to the helper T cells that in turn produce lymphokines. Thus idiotypic interactions amongst B cells require the concomitant activation of helper T cells (in our model we have in fact assumed that all B cells receive sufficient help). We are at present analysing the conditions for percolation in a network model with helper T cells [De Boer & Hogeweg, in prep.]. These preliminary results suggest that helper T cells hinder the development of idiotypic B-B cell interactions. Secondly, T cells might also form networks of idiotypic T-T cell interactions. Recent experimental data [25,26] strongly suggest that such interactions play an important role in immunoregulation. Several complications however hamper the development of reasonable models of such T cell networks: 1) are T-T interactions MHC restricted and, as a consequence, do these T cells express class II MHC which seems necessary for helper T cell interactions, and 2) do we have to incorporate a separate class of (the very controversial) suppressor T cells, or (which seems more interesting) can the suppressor phenotype be the result of the idiotypic interaction? Unfortunately, due to these open questions, these experimental data allow for the development of very many, seemingly equivalent, models. As yet, we conclude that the role that T cells play in the percolation of idiotypic B-B, B-T, or T-T signals remains to be established.

3. No functional network. The most interesting interpretation of our extensive percolation problem is that immune systems apparently do not function by means of such profound network structures. The existence of functional idiotypic interactions amongst "classical" B cells was also questioned by Eichmann [this volume]. Additionally, Urbain [27; this volume] proposed his "broken mirror hypothesis" which implies that idiotypic interactions occur in pairs and not in the form of profound networks. If idiotypic interactions were indeed to fail to form profound networks, the comparison with the powerful neural networks would break down, and we should no longer speak of immunoregulation by *network* properties. Even Jerne [10] speaks of "preferred partners" in idiotypic interactions. Also Cohn [28] questions the functional significance of idiotypic networks. Thus it seems quite reasonable to conclude that our percolation problem is a consequence of our erroneous assumption that immune systems function by means of profound network structures. Our preliminary results which show that helper T cells may hinder the development of functional idiotypic interactions would, if indeed immune systems do not function by means of profound network structures, provide a means of escaping from the seemingly inescapable idiotypic network theory [De Boer & Hogeweg, in prep.].

IgM networks. In empirical immunology, however, idiotypic network theory has roused considerable interest recently. Several levels of specific idiotypic interactions amongst monoclonal antibodies, i.e. "paper networks", have been described [29]. Moreover, it has been demonstrated that, in contrast to the adult immune system, B cells early in ontogeny have numerous idiotypic interactions [21,22]. Additionally, it has been suggested that the B cells from the early repertoire form a separate lineage of cells [30]. These IgM producing cell types most probably function independently of helper T cells and are hence indeed expected to form profound network structures [De Boer & Hogeweg, in prep.]. It is thus conceivable that, early in ontogeny, immune systems do form functional networks, and that these network properties disappear when the system matures. The possible significance, for e.g. self non-self discrimination, of such "early repertoire" networks seems an interesting point for further study.

Classical immune systems. For the classical, i.e. adult, immune systems that in fact display the phenomena of specific immune reactions, immunity, and tolerance, we apparently have to look for alternative modes of immunoregulation. One which remains close to idiotypic network theory is regulation by anti-idiotypic suppressor T cells [25,26]. Another, which seems more simple (and less controversial) is the clonal helper T cell response. The helper T cell proliferation process has, due to autocatalytic IL2 production, proliferation thresholds [31-33] that account for several tolerance phenomena, including self non-self discrimination. Finally, antigen presentation by macrophages and/or subsequently by B cells might play an important role in immunoregulation. We conclude that the clonal level of the immune system seems to provide sufficient immunoregulatory processes to account for complex "computation" or "information processing" .

5. Acknowledgements

This work has been done in close collaboration with Dr. P. Hogeweg, head of our Bioinformatics Group. I am grateful to Ms. S.M. McNab for linguistic advice.

6. References

1. N.K. Jerne: Ann. Immunol. (Inst. Pasteur) 125C, 373-389 (1974)
2. G.W. Hoffmann: J. Theor. Biol. 122, 33-67 (1986)
3. F.J. Varela, A. Coutinho, B. Dupire, N.N. Vaz: In Theoretical Immunology, ed. by A.S. Perelson, Part Two, SFI Studies in the Science of Complexity, Vol. III (Addison-Wesley, 1988) pp.. 359-375
4. J.J. Hopfield, D.W. Tank: Science 233, 625-633 (1986)
5. R.J. De Boer: In Theoretical Immunology, ed. by A.S. Perelson, Part Two, SFI Studies in the Science of Complexity, Vol. III (Addison-Wesley, 1988) pp. 265-289
6. R.J. De Boer, P. Hogeweg: Bull. Math. Biol. (in press)
7. R.J. De Boer, P. Hogeweg: Unreasonable Implications of Reasonable Idiotypic Network Assumptions (submitted)
8. G.W. Hoffmann: In Lecture Notes in Biomathematics, ed. by C. Bruni, G. Doria, G. Koch, R. Strom, Vol. 32 (Springer, Berlin, 1979) pp. 239-257
9. G.W. Hoffmann: In Contemp. Topics Immunobiol., ed. by N.L. Warner (Ed.), Vol.11 (Plenum Press, New York, 1980) pp. 185-226
10. N.K. Jerne: Immunol. Rev. 79, 5-24 (1984)
11. B. Goldstein: In Theoretical Immunology, ed. by A.S. Perelson, Part One, SFI Studies in the Science of Complexity, Vol. III (Addison-Wesley, 1988) pp. 3-40
12. J.D. Farmer, N.H. Packard, A.S. Perelson: Physica 22D, 187-204 (1986)
13. A.S. Perelson: In Theoretical Immunology, ed. by A.S. Perelson, Part Two, SFI Studies in the Science of Complexity, Vol. III (Addison-Wesley, 1988) pp. 377-401
14. GRIND: Great Integrator Differential Equations. R.J. De Boer, Bioinformatics Group, University of Utrecht, The Netherlands (1983)
15. B.A. Gottwald, G. Wanner: Computing 26, 355-360 (1981)
16. NAG: Numerical Algorithms Group. Mayfield House, 256 Banbury Road, Oxford OX2 7DE, United Kingdom (1984)
17. P. Erdos, A. Renyi: On the random graphs 1, vol. 6. (Debrecar, Inst. Math. Univ. DeBreceniens, Hungary 1959)
18. P. Erdos, A. Renyi: On the random graphs, Publ. No. 5, Math. Inst. Hung. Acad. Sci. (1960)
19. S.A. Kauffman: J. Theor. Biol. 119, 1-24 (1986)
20. J. Novotny, M. Handschumacher, R.E. Bruccoleri: Immunol. Today 8, 26-31 (1987)
21. D. Holmberg, S. Forsgen, F. Ivars, A. Coutinho: Eur. J. Immunol. 14, 435-441 (1984)
22. D. Holmberg, G. Wennerstrom, L. Andrade, A. Coutinho: Eur. J. Immunol. 16, 82-87 (1986)
23. L.A. Segel, A.S. Perelson: In Theoretical Immunology, ed. by A.S. Perelson, Part Two, SFI Studies in the Science of Complexity, Vol. III (Addison-Wesley, 1988) pp. 321-343
24. F. Melchers, J. Anderson: Ann. Rev. Immunol. 4, 13-36 (1986)
25. O. Lider, T. Reshef, E. Beraud, A. Ben-Nun, I.R. Cohen: Science 239, 181-183 (1988)
26. D. Sun, D., Y. Qin, J. Chluba, J.T. Epplen, H. Wekerle: Nature 332, 843-845 (1988)
27. J. Urbain: Ann. Immunol. (Inst. Pasteur) 137C, 57-64 (1986)
28. M. Cohn: Ann. Immunol. (Inst. Pasteur) 137C, 64-76 (1986)

29. M. Vakil, J.F. Kearny: Eur. J. Immunol. 16, 1151-1158 (1986)
30. K. Hayakawa, R. Hardy, M. Honda, L.A. Herzenberg, A.D. Steinberg, L.A. Herzenberg: Proc. Natl. Acad. Sci. USA 81, 2494-2498 (1984)
31. R.J. De Boer, P. Hogeweg: J. Theor. Biol. 120, 331-351 (1986)
32. R.J. De Boer, P. Hogeweg: J. Theor. Biol. 124, 343-369 (1987)
33. I.G. Kevrekidis, A.D. Zecha, A.S. Perelson: In Theoretical Immunology, ed. by A.S. Perelson, Part One, SFI Studies in the Science of Complexity, Vol. III (Addison-Wesley, 1988) pp. 167-197

The Concept of Idiotypic Network: Deficient or Premature?

Z. Grossman

Department of Physiology and Pharmacology, Sackler Faculty of Medicine, Tel Aviv University, Tel Aviv 69978, Israel and Pittsburgh Cancer Institute

"A discovery is premature if its implications cannot be connected by a series of simple logical steps to canonical, or generally accepted, knowledge" (1). If this is true, then the discovery of the idiotype-anti-idiotype network of lymphocytes (2) must be considered premature, since to this date it has not been successfully integrated, logically, into the growing body of knowledge about the immune system. Alternatively, the network hypothesis itself or the "accepted knowledge", or both, are deficient. I shall argue in favor of the second possibility. But first I shall briefly refer to the body of knowledge and discuss whether the analytical tools in the hands of theoretical immunologists are capable of integrating or reshaping it in any substantial way.

1. PERSPECTIVES OF IMMUNITY

The immune system obviously possesses both "descriptive" and "interactional" complexity, in the sense of Wimsatt (3). Such a system needs more than one perspective for its description.

1.1 The Specificity Perspective

The function of the immune system is traditionally envisioned as mediating immune responses against foreign (or modified) molecules and cells in order to destroy pathogenic agents. Pasteur demonstrated that attenuated organisms could be used as vaccines, to specifically protect animals against subsequent infection with the same virulent organisms. Understanding of the molecular, genetic and cellular basis of specificity has been the central goal of immunological science since then (4).

The "clonal selection" theory of Burnet (5) has been widely accepted as an explanation for the cellular basis of immunological specificity. Clonal selection is the selective activation by antigen and subsequent expansion of pre-existing B cell clones already specialized in production of complementary antibody to that antigen. The diversity of B cells and antibodies has been shown in the recent years to arise both from combinatorical association of distinct genetic elements in B cell progenitors and from somatic hypermutation. These principles have largely been extended to T lymphocytes. Similar combinatorial (but not mutational) genetic mechanisms generate the specificity and diversity of the T cell receptor.

In addition to their antigen specificity, another level of specificity is imparted to T cells by major histocompatibility complex (MHC) molecules ("MHC restriction"). Activation of T cells requires co-recognition of antigen and particular MHC-encoded molecules.

Finally, a major aspect of specificity is self-nonself discrimination, namely, the absence of agressive immune response of the body's lymphocytes against its own cells and molecules. Persistent deviations from this state lead to autoimmune disease.

Springer Series in Synergetics, Vol. 46 **Theories of Immune Networks**
Editor: H. Atlan and I.R. Cohen

1.2 The Developmental Perspective

Significantly, it was found that tolerance to self is not imprinted onto lymphocytes by a preprogrammed genetic event, but rather depends on intercellular interactions during their ontogeny. The same is true with regard to the acquisition of MHC restriction. According to a current model of T-lymphocyte ontogeny (6), maturation in the thymus occurs in two stages. In the first, receptor bearing immature thymocytes are positively selected by thymic ephithelial cells for weak binding to self MHC-molecules. In the second, potentially self-reactive cells that bind strongly to self-MHC molecules are eliminated in the thymic medulla through interactions with bone-marrow derived cells. The nature of the selective mechanism is unknown. In particular, it is difficult to imagine how cells could be selected for a binding affinity so weak as to be ineffectual in the absence of antigen. The presence of a hypothetical self protein has been invoked that helps to link T-cell receptors to MHC molecules on thymic epithelial cells, so as to make the overall binding affinity of the positively selected cells a strong one (6).

It is a common practice in the field to invoke, often on ad-hoc basis, specialized molecules and cells to account for observations at the cell population level. The key events both in development and during functional activity are seen as a series of signals delivered by specialized messangers. These messangers, their location and their mode of operation being the product of evolution, it is apparently believed that the respective events at different levels of development and function need not, as a rule, be comparable.

An alternative view is that the selectable objects for natural selection, besides the component messangers themselves, are the very principles of inter- and intracellular organization. Such principles cannot be entirely tissue-specific or stage-specific. In the present context, the mechanisms of selection which shape the T-cell repertoire in the thymus may also underly other immunologcial phenomena.

1.3 The Immunoregulation Perspective

A large array of cell types and their molecular products are coordinated in order to generate an immune response, including B and T lymphocytes, NK cells, macrophages and other hemopoietic cells, various lymphokines and other hormones and factors, and cell surface receptors for these factors and for antigens. In a recent presentation, Paul (4) described our knowledge of how the immune system actually functions as very primitive and predicted that immunological regulation will be the central intellectual problem of immunology during its second century. The foundation was provided by the work of Mitchison (7) and others on cooperation between helper T cells and B cells in antibody production; by the introduction of the concept of immunological suppression by Gershon (8); and by Jerne's (2) proposition that the immune system was regulated by a web of specificities interconnected by idiotypic-anti-idiotypic interactions. Given that lymphocytes are potentially coupled by cross-reactive recognition of antigens, by idiotypic interactions, and by the exchange of a large number of soluble factors, it becomes a major problem to explain why the responses are limited and highly selective in terms of magnitude, specificity and function (9).

There are no general rules for the quantitative prediction of the results of manipulation of the immune system: somehow, they should be determined by the balance between "helper" and "suppressor" cells of different kinds. Most of the existing schemes of regulation assign to each subset of cells a discrete, immutable function and a fixed relationship to other subsets. The dynamic aspect of regulation is then limited to variability in the number of cells in each subset. It is believed, that understanding of how the system works will follow if all such cells and their respective target cells could be identified. Clearly, this is an extrapolation of the genetic program idea. A "program of organization" is implied, imposing a functional structure. An alternative approach (10) will be discussed below.

A number of manifestations of quantitative regulation, in particular, deserve special attention. They can be divided, roughly, into three groups:

a. Manifestations of time-dependent selection. The classical example is "affinity maturation" - the gradual (or stepwise) increase with time in the average affinity of the antibody produced in response to antigenic challange.
b. Non-monotonic dose-response patterns. In experimental models, in vivo and in culture, the numbers of activated lymphocytes at the peak of the immune response do not grow monotonically with the amount of antigen which is introduced. Rather, the dose-response curves typically show a maximum at some intermediate dose. The peak value also depends on the rate of introduction, with gradual, slow rates usually generating weaker responses. Probably related manifestations are non-monotonic growth patterns of tumors such as the "sneaking through" phenomenon, where small amounts of transplanted tumor cells grow progressively, medium-sized amounts are rejected, and large ones break through again (11).
c. "Memory". Immunological memory and tolerance have been ascribed to long-lived cell populations, progeny of activated cells. Recent evidence, however, supports the view that individual B and T cells have a limited life-expectancy that cannot account on a per-cell basis for the longer persistence of memory (12). The evidence also indicates that peripheral T cells, but not B cells, have a considerable capacity to expand and suggests that continuous self-renewal (or its absence) of specific T cell clones may somehow be responsible for the maintenance of memory (or tolerance).

1.4 The Network Perspective

From this perspective, the specific recognition aspect of lymphocytes, considered always to be the central one, is also the key to the functional regulation of the immune system. The receptor molecules on lymphocyte clones, created by random genetic mechanisms, differ from each other not only at the recognition site, but also in a set of structures - "idiotopes", and collectively: "idiotype" - which serve as antigens for other clones. Thus each clone can activate some other clones, forming a network of interactions which could encompass the whole system. The functional consequence of such activation could be suppression, expansion, and/or induction of effector function in the clones involved, depending on the functional properties of the cells and perhaps also on the "direction" of the signal; no general rules are known. Idiotypic interactions have been demonstrated, but the functional significance is under debate (13, 10).

2. THEORETICAL MODELS OF IMMUNITY

Of the three measures by which models were evaluated, "realism, precision and generality" (14), the emphasis is necessarily on the first and third ones. It is generally impossible to make precise verifiable predictions when problems of "organized complexity" are encountered; that is, it is difficult to predict numbers, as distinct from "trends" (15). The purpose of a model here is to make valid generalizations with regards to causality as it operates in a complex biological system (16).

The essential part of a model is a hypothesis, stated verbally and pictorially; a mathematical representation is often secondary and optional. The relationship between hypotheses and their analyses, whether quantitative or qualitative, is essentially the same as that between hypotheses and experiments as defined by Medawar (17, 16). There are three kinds of experiments/analyses:

1. Demonstrative or Aristotelian. Intended to illustrate a preconceived truth and to convince others of its validity.
2. Critical or Galilean. Actions carried out to test "a hypothesis or preconceived opinion by examining the logical consequences of holding it".
3. Deductive or Kantian. Examination of the consequences of varying the axioms or presuppositions of a scheme of deductive reasoning.

2.1 The Demonstrative Approach

Commenting on the first category, Humphrey had only this to say, that such experiments are "the stuff of a good many Ph.D. and M.D. theses, which fill the libraries of universities and the pages of journals without adding greatly to scientific knowledge". Mathematical models of this category deserve, perhaps, a more favorable evaluation. First, the hypotheses may include nonlinear interactions, entailing "counter-intuitive" causal relationships among observables and requiring "illustration" and "convincing". Furthermore, a mathematical model can be used, to some extent, to demonstrate or refute the self-consistency of the statements made in the hypothesis. For instance, the hypothesis may require that certain types of steady states exist and be stable. While a complete stability analysis of high-dimensional systems presents substantial technical problems, some attempts at qualitative or partial configuration stability testing in immunology have been described (18, 19, 15).

A recent model of autoimmunity (20), involving anti-idiotypic interactions (but not a real network in the sense of Jerne), is another example. Four classes of regulatory T lymphocytes were earlier implicated in the control of experimental autoimmune diseases: a pair of antigen-specific helper and suppressor T lymphocytes that recognize the self-antigen; and a pair of anti-idiotypic helper and suppressor T lymphocytes that recognize the autoimmune effector lymphocytes. The hypothesis is that "non-immunogenic" self-antigen activates only suppressor cells; that immunogenic self-antigen activates both helper and suppressor cells; that (activated) helper cells alone can activate autoimmune effector cells, but the combination of both helper and suppressor cells leads to suppression; and that (activated) effector cells activate anti-idiotypic helper and suppressor cells. It follows directly from these presuppositions that, other things being equal (that is, if no additional factors are invoked which influence the help-suppression balance), the effector cells are likely to be suppressed even in the presence of immunogenic antigen; and that removal (or weakening) of one of the two suppressor cell populations may allow a transition to an activated state of the effector cell population. These are consequenses of the subtle functional dominance of suppressor cells over helper cells. The additional assumptions (20) are not only ad hoc but also far-fetched biologically. Each of the five populations of cells is represented as an automaton having two states, "on" and "off". It is hard to imagine, even as an approximation, how real cell populations could simultaneously switch, at random, between collectively activated and non-activated states. Nevertheless, the model represents the relevant features of the hypothesis and is thus a legitimate abstract analog. The numerical simulations lead to the expected behavior. However, the analysis does not "suggest how self-tolerance may operate, how progressive autoimmune disease may develop, and how T cell vaccination can control autoimmune disease" (20); rather, it demonstrates in a symbolic (or metaphoric) way the authors' pre-conceived ideas about these issues.

2.2 The "Critical" Approach

Most mathematical models belong, by their builders' claims, to Medawar's second category. The researchers formulate a hypothesis - however limited - write differential equations and study the solutions as the "logical consequences" of the hypothesis. Then they look for consistency with the observed behavior and sometimes propose additional experiments to test the "predictions of the model".

There are several pitfalls in this procedure. Some have to do with the notion of "prediction". For the sake of mathematical completeness, and also because our knowledge of the system is only partial, ad hoc assumptions are included in the hypothesis (often "for simplicity"). To the extent that the results of the model's analysis depend on such assumptions they should not be regarded as predictions and can hardly motivate experiments to test them. Other results may be found that depend on more robust aspects of the hypothesis, such as the relative specificities of stimulation and suppression, the degree of cooperativity at various levels of interaction, the conditions for cellular proliferation versus differentiation, and general structural properties of a network. The results are then legitimately presented as predictions of the model, but this may still be somewhat misleading. In fact, the qualitative aspects of the results can often be inferred directly from the assumptions, without detailed analysis, and the quantitative details, such as the parametric range for which they hold, are of little significance anyway. Again, the mathematical model, with its more severe assumptions, serves mainly to illustrate and support qualitative statements of the hypothesis and therefore belong in the first category. The secondary role of such mathematical models should be acknowledged so as not to mask the real issue of "valid generalization" about causality with artifacts of the mathematical methodology.

Other pitfalls stem from the inadequacy of a mathematical formulation as such to deal with the issues of realism and generality. In some models which describe relatively small systems of cells and molecules, a number of differential equations are written on the basis of the current phenomenology complemented with ad hoc assumptions. Using available algorithms based, e.g., on bifurcation theory, a numerical search in parameter space is performed to identify phase space singularities and their local stability as a function of some parameters. These singularities define qualitatively different behaviors of the nonlinear model that are believed to be analogous to experimentally observed ones. The models are admittedly "caricatures of the in vitro and in vivo situation"; yet the authors may believe that they "predict behaviors that are immunologically sensible" (21). Other models, more comprehensive and detailed, attempt to be "realistic" about certain aspects, such as patterns of receptor cross-reactivity, binding kinetics and the detailed book-keeping of a large (and variable) number of clones. On the other hand, other aspects are ignored or treated simplistically, notably, cellular cooperation, proliferation versus differentiation, and various feedback control mechanisms. It is believed that new elements and new regulatory pathways can be added to, or included in, existing models in stages, one at a time, to make them more realistic.

This view, that phenomenological knowledge can be assembled and put together gradually into nonlinear differential equations which will eventually reveal the general order of things, is an extension of the programmatic approach in experimental biology, as was noted earlier: it implies a "program of organization". Resembling the "aufbau" method in atomic physics (22), its philosophical basis can perhaps be found in the conjecture that hierarchically organized biological complexes (molecules, cells, tissues, organisms) have evolved by means of an aufbau process. The utmost justification of models based on this approach should rest on their success to make valid nontrivial generalizations about the systems being modeled and beyond. Such success can hardly be claimed as yet. In particular, even comprehensive idiotypic network models that have been constructed do not advance our understanding of how the network can regulate itself and, if it has functional significance, what functions it performs and how. Functionally, a strongly coupled network indeed appears to be "tied in a Gordian knot" (23).

One such model has recently been proposed by Perelson et al (24, 25). Following the customery method of pattern matching (e.g., in neural networks), antibodies, B cells and antigens are represented by bit strings, and the degree of complementarity of any pair of strings is quantified using ad hoc rules and associated with binding affinity. A random algorithm is used to generate such strings, forming multiply connected groups. Given such a system of strings, differential equations can be written to describe the turnover of cells and molecules. One version (25) includes bilinear idiotypic stimulation and suppression terms (depending on the "direction" of recognition) which are proportional to the affinities; an algorithm to generate random mutations (i.e., new strings); and an algorithm to update the list of variables. More general equations have also been constructed in which the chemistry of heterogeneous mixtures of antibodies, B cell receptors and antigen leads to more sophisticated interaction terms. However, the basic issues remain unresolved. The dubious advantage of triggerring many clones against a complex antigen through "vertical" as well as "horizontal" connections is offset by potential violation of specificity and by noise. One generalization that Perelson has associated with his network model (24) is the proposition that immunological memory may be dynamic, that is, actively maintained through co-stimulation of idiotypically related clones. However, this proposition is not new and the notion does not depend on this model's particular formulation or simulation. Memory has been associated with the multiplicity of stable steady states (9, 10; see below). In particular, it has been proposed that idiotypic co-stimulation could dynamically support the maintenance of antigen-speficic memory even in the absence of residual antigen (10). I suspect that, beyond these generalizations about dynamic memory, unless new perspectives are introduced, we shall learn from simulations of particular ad hoc models - comprehensive as they may be - more about the models themselves than about the biological phenomenon.

In summary, in what regards mathematical models of immunity, the "Galilean" approach is either not applicable or not critical enough. Most of these models represent a non-critical attitude towards the established presuppositions of experimental biologists, which tend to favor highly structured and preprogrammed cellular communication. This limits the potential of the models to suggest radically new interpretations of the observed phenomena. The hope that additional insights will "emerge" from mathematical analysis of such models seldom comes true.

2.3 Deductive Reasoning

Models of the third category are scarce. This may be due to the reluctance of most model builders to depart from the unidirectional road from observation to theory and indulge in "speculative" thought (26).

As Humphrey (16) stresses, an attempt to turn currently accepted hypotheses upside down is likely to be worthwhile only if there are indications that these hypotheses are deficient or susceptible to modification. My own approach developed with a number of premises in mind.

I. The concepts of structural specificity (clonal selection), pre-programmed functional differentiation, and regulation by helper/suppressor cells or by idiotypic recognition network are too rigid and too limited to coherently explain the observed phenomena. There are many discrepancies and many ad hoc explanations.

II. While the above concepts each represents a different perspective of the immune system as if they were independent of each other and of operational principles in other cell systems, it is worthwhile to consider a more holistic scheme. In that scheme development, regulation and the expression of specificity will appear (operationally) interrelated in that they depend on common cellular mechanisms of general physiological significance.

III. New strategies of self-organization should be explored to account for adaptability and coordination in the immune system.

The first key step was to introduce the concept of "balance of growth" into regulation schemes (27, 28). A decade ago, and largely also to this time, it had generally been assumed that division and differentiation of a triggered lymphocyte (or other mitotic cell) follow in a pre-programmed way. Consequently, the focus of interest was on the events leading to binding of antigen, or "recognition". Contrary to some theories, however, there is no obvious causal relationship between division and differentiation-maturation at the single cell level. We assumed that initiation of differentiation and entering into mitosis were competing cellular events, and that the relative probabilities were regulated by extracellular signals, including antigen binding. An additional distinction was made between "resting" and "activated" cells: only activated cells can be further induced to mature or regenerate. Differentiation occurs in several steps and is eventually accompanied by impairment of the capacity for self-renewal. We assumed that this holds for T lymphocytes as well as for other cell types, proposing that differentiation is antagonistic to clonal expansion as a general rule. This scheme suggests that shifting the balance of growth from proliferation to differentiation could be a major control mechanism of an ongoing immune response, even in the absence of other negative feedback mechanisms that can be conceived.

It can readily explain the non-monotonic dose-response dependence (28). Proliferation thresholds are predicted in terms of antigen quantity or affinity, defining low-zone and high-zone regimens of low-responsiveness. Moreover, antigen provides both direct and indirect signals. The delayed accumulation of activated helper cells, in particular, could amplify the direct antigenic effects and in this way shift the balance of growth and control the response. At the cell population level they appear to switch functions, from help to suppression. Several situations involving responder and helper clones, stimulated simultaneously or sequentially and under different conditions, have been qualitatively analyzed and interpreted (10).

The second major step in developing the theory was revision of clonal selection. Burnet's clonal selection is actually clonal amplification based on structural compatibility between molecules. In the revised form it is a true dynamic selection, determined by the relative growth rates of competing lymphocytes (9, 10, 27). The non-monotonic dependence of lymphocyte growth rates on antigen dose and affinity, and the interplay of several kinds of stimuli and suppressive mechanisms, endow the system with a considerable flexibility and allow us to consider a broader range of phenomena and identify general strategies of selection which have adaptive value, namely, functional significance.

The functions accounted for are: focusing of an initially disperse immune response, maintenance of specific memory or tolerance, self-nonself discrimination, and acquisition of MHC-restricted responsiveness. Collectively this amounts to a dynamic "regulation of specificity" (10), integrating these two perspectives of immunity together. Interactions of lymphocytes, at different stages of differentiation, with foreign antigens and with the familiar molecular and cellular environment are subject to the same selection mechanisms, relating development to functional immunity and allowing inferences to be made in both directions.

Response focusing was accounted for by postulating that suppressive mechanisms are generally less specific than activation (9, 10). This proposition has gained considerable experimental support. Cross-reactive feedback control defines a competitive relationship among responding cell populations and can lead to clonal dominance or to other patterns of selective clonal expression, depending on the particular patterns of cross-reactivity and on the duration and strength of stimulation. It was estimated that relatively small differences in affinity, for example, can in weeks or even days have substantial effect on the size of the dominant clone (27). Complex patterns of affinity maturation can be envisioned when cross reactivity exists both at the level of activation and of suppression (9, 15). The combination of differential stimulation and less specific, or non-specific control could lead to negative selection of specific clones as effectively as can specific suppresion. This consideration, and also the proposition that helper T lymphocytes as a population exhibit a dual function, help or suppression, suggest that specialized suppressor T lymphocytes may be redundant or non-essential (10).

An important mode of selection predicted by the hypothesis involved "latent proliferation": proliferation of lymphocytes with little or no expression of effector function (9). I proposed that, in addition to the dynamic proliferation thresholds mentioned earlier, activation, self-renewal and differentiation were threshold-dependent in terms of affinity also at the individual cell level, and that the differentiation threshold was generally higher. This implies that proliferation can be uncoupled from differentiation under certain predictable conditions. As a rule, in the steady state mode of interaction of lymphocytes with antigen and with their environment, proliferation should be essentially latent: as, by hypothesis, the induction of terminal differentiation is growth-inhibiting, and because growth leads to prominence, differentiation and expression of effector function are dynamically restricted.

Dynamic selection tuned by the balance of growth with latent proliferation as its likely consequence offers a strategy for self-nonself discrimination (9). Referring to T cells in particular, those clones will gain prominence during maturation which on the one hand are induced by self recognition into extensive proliferation and, on the other hand, resist environmental pressures to undergo premature terminal differentiation. Other cells will be eliminated. It appears that subsets of T lymphocytes that interact with self antigen presented via particular MHC-coded molecules on thymic epithelial and hemopoietic cells possess this favorable balance of growth. Hence, in subsequent encounters, the selected T lymphocytes interact with other cells in an MHC-restricted way. Recognition of self continues to be of functional importance (9, 29), but it does not lead to agression against self, either because the affinity is too low or because signal transduction pathways are so tuned as to restrict the outcome (9). Recognition of cells bearing foreign antigens in association with MHC molecules by some of the selected lymphocytes will lead to further differentiation and expression of effector function, either due to an upward shift in the affinity of binding or because of a change at the level of transduction.

The distinction of non-self from self, especially in a primary response, depends also on a kinetic aspect, namely, the rate of introduction of antigen. The magnitude of a response depends not on the size of the perturbation per se, but rather on its time derivative. Clones of relatively high affinity usually acquire selective advantage during the first stage following an abrupt perturbation due to their larger rates of activation and self-renewal. Time delays in the differentiation of precursor lymphocytes into non-dividing effectors and in the induction of effective feedback suppression allow the response to "overshoot". This in turn can lead to a cyclic response (28) and - more importantly - may be essential for the extinction of a pathogenic agent. The effect was demonstrated in a model describing the interaction of an experimental tumor with a clone of cytotoxic T cells (11). The model exhibits the "sneaking through" phenomenon: intermediate numbers of transplanted tumor cells, beyond the dynamic proliferation threshold, generate overshooting response and subsequent tumor elimination; a too large or - significantly - too small initial tumor size leads to tolerance and to tumor escape from immune elimination.

The excessive accumulation of responding cells following a transient antigenic perturbation is a decisive factor also in the maintenance of a specific memory state. Cooperative (or autocatalytic) effects within the responding cell population tend to stabilize a sufficiently expanded population at an elevated state, while competing non-expanded populations are stably suppressed. Persistence of residual antigen, mediating cooperation among antigen specific subsets of lymphocytes ("help"); idiotypic co-stimulation; or simply the localized nature of the co-stimulatory or self-stimulatory effects, could provide for the required specificity of this dynamic memory. If the original antigen selects for a latent cell population, similar mechanisms can stabilize the dominance of that population; in this case a memory of non-responsiveness - namely, tolerance - is dynamically maintained.

2.4 The Network is Flat!

Having outlined a theoretical scheme in which immunological specificity and function are dynamically regulated at the level of cell populations - thus linking together specificity, regulation and development - there remained no room for an autonomic concept of idiotypic network playing a unique regulatory role (9, 10). Formally, self interactions via inter-receptor recognition, as well as recognition of MHC and other molecules, generate an infinitely coupled network. On the basis of the theoretical scheme and of existing evidence. I proposed that, functionally, this network is "flat" - connectivity is severely constrained by the rules of dynamic selection. In the absence of external antigenic stimulation idiotypic interactions are latent, in the sense defined, characterized by low affinity and minimal expression of effector function. Recognition of antigen-modified receptors could transiently elicit effector responses, including the cooperative or suppressive effects attributed to anti-idiotypic regulatory cells. Such antigen-restricted response necessarily involves only a few clonotypes. Also, because of the dependence of the immune response on perturbation rates, if cell populations trigger each other sequentially, a quickly decaying response along the chain due to attenuation by feedback forces is predicted (10). Latent idiotypic interactions could drive cell replication and affect the specificity repertoire. In particular, idiotypic cooperation can support the maintenance of antigen-specific memory in the absence of residual antigen (10).

Deductive reasoning is followed also by Segel and Perelson (35) and by De Boer and Hogeweg (36) in examining the question of connectivity. These authors formulated highly simplified mathematical models, aiming "to obtain an overview of the situation and hence to pose strong general questions" (35). The conclusions are similar to mine. Highly-connected idiotypic networks fail to account for memory phenomena and for the control of proliferation (36). If recognition structures on lymphocyte receptors are characterized by a single continuous variable (35), with a simple rule to quantitate receptor-receptor binding affinity, "the parameters could be such that network interactions are not at all important". These parameters probably represent weak idiotypic coupling (in the absence of antigen) - implying a high proliferation threshold - yielding vigorous local response to a sudden perturbation and strong attenuation along the network. Interestingly, requiring that the ground state with a uniform distribution of clonal sizes should not be too stable, so that it could be excited into different size-distributions, the model (35) suggests that activating influences are more specific than suppressive ones, as I proposed (9; see above). Although the context is different, I do not think that the results are entirely independent: dynamic selection and non-uniformity seem to be logically related.

2.5 Some Other Recent Developments

In a series of papers, De Boer and Hogeweg (30, 31, 32) presented models of T lymphocytes and discussed immunological tolerance and self-nonself discrimination. They have clearly adopted the "balance of growth" principle, though not the same balance of growth configurations which I proposed (A rather sketchy comparison can be found in 30).

The essence of the balance of growth concept is that increased stimulation shifts the balance towards differentiation into short-lived and/or growth impaired effector cells (9). I proposed this as a major control mechanism in different cell systems (33), noting that it can obviate the need for negative control mechanisms. To specify the precise way in which proliferation and differentiation are related, one has to make additional assumptions. In the configuration which I proposed differentiation occurs in several steps and in each step a resting cell is activated and then it either self-renews or differentiates further. The terminal differentiation state is short-lived and/or non-mitotic (see 27, Scheme 2). The resting states at different stages of differentiation represent precursor and memory cells; however, a single-state configuration is sufficient to demonstrate the properties of the model qualitatively. The per-cell activation and differentiation rates were assumed to depend linearly on antigen, and this was sufficient to produce two proliferation thresholds: the lower one essentially where proliferation supersedes the unperturbed turnover, and the higher where differentiation exceeds proliferation. Dependence on helper signals, whether self-induced or via a helper cell population, was also examined qualitatively (27, 9, 10); it was not considered essential in a single-clone model.

De Boer and Hogeweg's configurations are different, structurally, in only one aspect of significance: the short-lived effector cells are also identified as the proliferating cells (roughly, my Z cells and the most differentiated Y cells in scheme 1 or scheme 2, ref. 27, become identical). The parameters of the model were assumed to be such that significant proliferation required clonal expansion by other means first. Thus the balance of growth was determined essentially by competition between cell accumulation and cell differentiation. Accumulation is facilitated by assuming that the immediate precursors of the differentiated effectors have long life-times (months, years). Strong antigenic stimulation induces this precursor/memory pool into a short term effector activity. The positive thymic selection of low-affinity, self-recognizing T cells could now be described as "latent accumulation" rather than "latent proliferation"; otherwise, the interpretation of self-tolerance induction and MHC-restriction remains the same. The authors assumed that the proliferation of effector T cells is autocatalytic; this facilitates transient proliferation and a more vigorous response when large precursor/memory clones are strongly activated. It should be noted that clonal expansion by proliferation is secondary here to expansion by accumulation.

From this comparison it can be concluded that the interpretation of phenomena related to tolerance and to self-nonself discrimination is facilitated, with no recourse to suppressor cells, by invoking any one of several models that realize the balance of growth concept. The formulation of detailed mathematical models and their numerical analysis may sometimes mask rather than clarify this deductive relationship. Different scenarios that can be envisioned should be discriminated on biological grounds. The existence of long-lived memory cells have been questioned (12). In addition, extensive (latent) proliferation appears to be involved in the selection of the T cell repertoire (12, 34).

Recently, some key elements of my dynamic selection hypothesis (9, 10) have been "rediscovered" and its empirical basis extended (37). These elements include competition among cell populations in a limiting environment - in particular competition for growth factors (IL-2); "specific induction and nonspecific expression of suppression"; the redundancy of specialized suppressor cells "with both specific function and a definite phenotype"; suppression by competition with expanding cell population whose "effector function is not what is being quantitated" (i.e., latent proliferation as the hidden cause for suppression); and the significance of the feature of excessive immunologic response (overshooting).

Other predictions of the hypothesis, the helper-suppressor duality (10) and the regulatory role of self recognition (9, 10) via MHC-coded or MHC-restricted molecules beyond thymic differentiation, have gained strong support in recent studies (38, 39, 40). Self-reactive T cells could amplify, suppress or contrasuppress T cell help, depending on the type and relative ratios of self and antigen-reactive cells. The opposing regulatory functions are related to modulation of the balance between proliferation and differentiation of activated B and T cells and need not be attributed to the existence of discrete T cell subsets (38).

Thus, "immunoregulation is a dynamic property of all cell populations which can perform variable functions depending on the concurrent activity of other interacting lymphocytes". Other studies have also documented the multiplicity of regulatory potential by T cell clones (e.g. 41). Self-reactive T cells provide help for the generation of cytotoxic cells with broad anti-self specificity (40). "The known bias of T cells for self-MHC is important not only in antigen recognition but also appears to be essential for immune regulation" (39, 40). All these findings have been anticipated in my theoretical analyses.

3. TOWARDS A THEORY OF ADAPTIVE NETWORKS

Variable probabilities of cell division and differentiation and cellular cooperativity and competition endow the immune system with a capacity for adapting to the self environment (tolerance) and to antigenic perturbations (focusing, memory). But evidence has been accumulating that indicates additional levels of complexity and suggest greater adaptability and new forms of self organization. The observations include the breakdown of the concept of single factor-single purpose specificity for lymphokines (4), which turn out each to be (a) multifunctional and (b) "promiscuous" in terms of the range of target cells; evidence for multifactorial cell growth control (see 42); manifestations of associative recognition and the need for "accessory" signals in lymphocyte activation; in vivo clustering of different lymphocytes and accessory cells (43); the existence of alternative pathways of lymphocyte activation (44); and the broad spectrum of bidirectional communication between "specialized" cells belonging to the immune, hemopoietic and neuroendocrine systems (45).

There is a growing schism between the traditional concept of antigen-oriented lymphocytes that function as the specialized members of a defence system vis-a-vis this unexpected biochemical diversity and degeneracy of intercellular signalling and intracellular signal processing. Why should lymphocytes, with their clonally distributed receptors for antigens, require so rich and so complicated system of communication with themselves and with other cells? How can multifunctional factors mediate precise responses? Unless we accept the radical reductionist approach of mapping functions directly to chemical activities, we are compelled to consider the possibility that the cells "compute" something, as individual units or in groups, and to ask what is the goal of the computation.

3.1 Signal Pattern Specificity and Learning

It has been suggested that lymphocytes have general regulatory functions beside their role in classic immune defense (46, 47, 29). Our own hypothesis was that lymphoid cells are involved in forcing and steering the differentiation of several types of cells (29). It is then conceivable that even cells outside the nervous system are required to be able to deal adaptively with rich classificatory challenges in perception of their environment, richer than usually appreciated. The hierarchical coordination of activities in tissues and organs may require that cells respond to biochemical signals in a context-dependent way, and that they tune and update their responsiveness to achieve appropriate discriminatory capacity.

These functional considerations and the evidence which I have outlined suggested a new scheme of self-organization (48, 49). Because of space limitations I shall only list the main points here:

1. Cells of immunologic and hemopoietic origin respond (in vivo) preferentially to particular combinations or arrays of signals.
2. Proliferation and semi-permanent or permanent changes in the pattern of gene expression - and perhaps even gene rearrangements - are adaptive cellular responses to perturbations which affect key metabolic activities of the cell.
3. Through such adaptive activity cells learn to respond preferentially to recurrent combinations of signals (evaluated over time).
4. The functional unit is a heterogeneous group of turning-over cells. Under stationary conditions, the group (but not necessarily individual cells) will maintain a stable (but resilient) phenotypic profile.
5. When lymphocytes (or their precursors) migrate into totally unfamiliar sites they are subject to a process of selection. Some positively selected cells eventually engage in latent, non-agressive co-stimulation among themselves and with other resident cells ("tolerance").
6. Antigenic perturbation leads, transiently at least, to a modified association of activated cells and signals. Some lymphocytes establish enhanced responsiveness to signals exchanged with other coactivated cells under the selective influence of the antigen. Memory designates the maintenance, through repeated interactions, of the new hierarchy of cells and signals even after disappearance of the antigen.
7. Immunologic agression appears during antigenic perturbation as a (usually transient) manifestation of mal-adaptation.

The "networks" which begin to emerge from these conjectures and speculations have little resemblence to idiotypic networks. They are groups of interacting cells, assembled on a somewhat ad hoc basis under the selective influence of patterned external perturbations. The communication among the cells in the group is only partially and variably dependent on specific recognition via the B and T cell receptors for self antigen; it requires coordinated expression of particular sets of interaction pathways (receptors, messanger molecules, etc.) which are not immunologically specific in the usual sense. The cells must be physically associated together, at least part of the time, but the association may be quite loose and unstable. The identity of cells with respect to the network hierarchy is actively maintained by latent mutual stimulation of cells belonging to the network. External perturbations are essentially ignored if they do not sufficiently resemble the stationary patterns which keep the network together; or, if they do, they trigger a hierarchy of adaptive responses.

3.2 Neural Versus Immunological Networks

That similarities exist in the mechanisms underlying immunity and brain function was suggested by Jerne (50, 51) and Cohn (52). Analogies drawn between the immune and central nervous system to-date suffer, in my view, from two serious drawbacks. The first one is an over-emphasis of one particular theoretical paradigm, the idiotypic network (51, 53-55). The trend is understandable, since this theory pictures the immune system as a highly connected system of simple units with well-defined connections of variable strengths and with simple modes of interaction. Such a system resembles popular versions of neural networks. The analogy as well as the network theory itself, however, fall short of being able to provide useful interpretation of immunological phenomena.

The second and more fundamental drawback is the failure to propose a real parallelism in the functional domain between the two systems. The task assigned to the immune system, to eliminate from the body anything that is perceived as being non-self, is hardly comparable to any of the brain functions and can be explained, essentially, in terms of Burnet's clonal selection. Describing the binding of receptors to molecules with complementary shapes as "pattern recognition", or comparing "immunological memory" of previous antigenic challenge, based on the maintenance of expanded clones of lymphocytes, to memory as it operates in the brain, are nothing but colorful metaphors.

There is a basic asymmetry in the problems facing us in trying to decipher the operation of the brain and of the immune system. In the case of the brain, there is enormous structural and chemical complexity, presumably designed to match the enormously complex functions of hierarchical pattern perception, associative memory, learning, coordination, etc. In the case of the immune system, we discover an increasing degree of biochemical and organizational complexity, not knowing its purpose and being unaware of the existence of a large range of functions to match. Relating this complexity to function is the key problem.

Could other tissues too conform to some of the principles of organization-related performance that presumably govern brain function? I have attempted to indicate both the need for previously unexpected capacities for signal classification and learning in such tissues and ways in which these capacities can be achieved, thus laying the ground for a genuine analogy. As Edelman noted (56), enormous epigenetic variation occurs at the level of membrane receptors and channels, secretory activities, axoplasmatic flows, and intracellular structures and processes, in lymphocytes and hemopoietic cells as well as in brain cells. Much of this variation is probably related to subtle functional and cognitive differences and forms the basis for adaptive self organization. I have proposed that cells of all kinds are not entirely programmed to respond to signals, that is, to "information", but also participate in the definition of information while sensing each other's activities. The strategy which I favor is feature discovery by competitive learning (see 57, 58). The goal is to categorize biochemical activity in the environment in terms of the most regular patterns and to generate, dynamically, a phenotypic mapping of those. Such mapping is envisioned as embedded in groups of cells manifesting coordinated and interdependent interactions. These groups can also store the memory of previously existing sources of prominent biochemical activity. Tracing recurrent patterns by way of shifting pattern specificities allows the system to anticipate the "meaningful" fluctuations around it and to readjust efficiently. In contrast to brain cells, other cells at different stages of adaptation self-renew and this provides additional plasticity to proliferative systems: although individual cells become more and more committed in terms of cognition and function as they mature, the whole population is never irreversibly committed and can accomodate modified or new patterns. Obviously, the perceptive capacities ascribed to turning-over cells are comparable only to relatively lower cognitive functions of the brain. Interestingly, Edelman's group selection theory of brain function (56) was inspired by the idea of clonal selection in immunology. He did not apply the theory to the immune system because, as he later noted, "cross-reactivity does not appear to be generally useful in the immune system, whereas cross-recognition can be highly important in an adaptive neural system" (59). This reservation may perhaps have to be reassessed now, if greater adaptive capacity and broader scope of functions are to be assigned to lymphocytes and to other cell types (60-62).

This overview suggests a program for further research. Biologically sensible algorithms pertaining to turning-over cells are to be formulated. Constructing such algorithms should be guided by implementational (micro-physiological) considerations as well as by a formal, top-down definition of the problem that the system is aiming to solve (63). After such algorithms are constructed, they may be tested for their self-consistency, stability, and for their capacity to categorize patterns, using mathematical analysis and simulations. Finally, experimental paradigms must be suggested to assess hypotheses about the major issues: to what extent can a cell such as a lymphocyte be viewed as an adaptive system? How adaptable is the organization of the tissue, under normal or perturbed conditions? Can pathological processes related to autoimmunity or other immune deficiencies, or to cancer progression, be explained as cognitive aberrations?

4. CONCLUSION

Immunologists used to ascribe to the immune system largely exclusive modes of activity, not shared by most other cell types. The idiotypic network is an example. I have stressed the importance of understanding the ways in which general rules of cell activation, organization and growth permit diverse cell populations to carry out their unique specialized functions.

Present models of the immune system, and of idiotypic networks in particular, are inherently limited in their ability to resolve basic questions. I suggest a more critical assessment of accepted notions about the nature of the immune system, including its postulated functions and the concept of specificity. Within the proposed theoretical framework there appears to be little justification for an autonomic concept of idiotypic network playing a unique regulatory role.

REFERENCES

1. G.S. Stent: Sci. Am. 227, 84, (1972)
2. N.K. Jerne: Ann. Immunol. (Paris) 125C, 373 (1974)
3. W.C. Wimsatt: In Topics of the Philosophy of Biology, ed. by M. Grene and E. Mendelsohn (Reidel Publ. Co., Dordrecht and Boston, 1976)
4. W.E. Paul: J. Immunol. 139, 1 (1987)
5. F.M. Burnet: The Clonal Selection Theory of Acquired Immunity. (Univ. Press, Cambridge, 1959)
6. P. Marrack, I. Kappler: Science 238, 1073 (1987)
7. N.A. Mitchison: In Cold Spring Harbor Symp. Quant. Biol. 32, 431 (1967)
8. R.K. Gershon, K. Kondo: Immunol. 18, 723 (1970)
9. Z. Grossman: Eur. J. Immunol. 12, 747 (1982)
10. Z. Grossman: Immunol. Rev. 79, 119 (1984)
11. Z. Grossman, G. Berke: J. Theor. Biol. 83, 267 (1980)
12. A.A. Freitas, B. Rocha, A.A. Coutinho: Immunol. Rev. 91, 5 (1986)
13. M. Cohn: Cell. Immunol. 61, 425 (1981)
14. R. Levins: Am. Sci. 54, 421 (1966)
15. Z. Grossman: Lec. Notes Biomath. 57, 312 (1985)
16. J.H. Humphrey: Ann. Rev. Immunol. 2, 1 (1984)
17. P.B. Medawar: Induction and Intuition in Scientific Thought (Methuen, London, 1969)
18. J. Eisenfeld: Ann. N.Y. Acad. Sci. 504, 132 (1987)
19. J. Eisenfeld, C. DeLisi: In Mathematics and Computers in Biomedical Applications, ed. by J. Eisenfeld and C. DeLisi (Elsvier, New York, 1985) p. 39
20. I.R. Cohen, H. Atlan: Network regulation of autoimmunity: an automaton model. Preprint.
21. I.G. Kevrekidis, A.P. Zecha, A.S. Perelson: In Theoretical Immunology, Part One, SFI Studies in the Sciences of Complexity, ed. by A.S. Perelson (Addison-Wesley, 1988) p. 167.
22. A.R. Peacocke: An Introduction to the Physical Chemistry of Biological Organization (Clarendon Press, Oxford, 1983) p. 13
23. R.E. Langman, M. Cohn: Immunol. Today 7, 100 (1986)
24. A.S. Perelson: In Theoretical Immunology, Part Two, SFI Studies in the Sciences of Complexity, ed. by A.S. Perelson (Addison-Wesley, 1988) p.
25. J.D. Farmer, S.A. Kaufman, N.H. Packard, A.S. Perelson: Ann. N.Y. Acad. Sci. 504, 118 (1987)
26. W.M. Elsasser: Reflection on a Theory of Organisms (Orbis, Quebec, 1987)
27. Z. Grossman, I.R. Cohen: Eur. J. Immunol. 10, 633 (1980)
28. Z. Grossman, R. Asofsky, C. DeLisi: J. Theor. Biol. 84, 49 (1980)
29. Z. Grossman, R.B. Herberman: Immunol. Today 7, 128 (1986)
30. R.J. De Boer, P. Hogeweg: IMA J. Math. Applied in Medicine and Biology 4, 1 (1987)
31. R.J. De Boer, P. Hogeweg: J. Theor. Biol. 124, 343 (1987)

32. R.J. De Boer, P. Hogeweg: J. Theor. Biol. 120, 331 (1986)
33. Z. Grossman: In Math. Modeling in Science and Technology, ed. by A. Avula et al. (Pergamon Press, New York, 1984) p. 933
34. C. Penit: J. Immunol. 137, 2115 (1986)
35. L.A. Segel, A.S. Perelson: In Theoretical Immunology, Part Two, SFI Studies in the Sciences of Complexity, ed. by A.S. Perelson (Addison-Wesley, 1988) p. 321
36. R.J. De Boer, P. Hogeweg: Symmetric idiotypic networks. II. Stability and unresponsiveness in high-dimensional models. Preprint
37. P.H. Sugarbaker, et al.: Cancer Bull. 39, 38 (1987)
38. J. Quintans, et al.: J. Immunol. 136, 1974 (1986)
39. H. Suzuki, et al.: J. Mol. Cell Immunol. 2, 331 (1986)
40. H. Suzuki, J. Quintans: J. Mol. Cell Immunol. 2, 345 (1986)
41. C. Clayberger, R.H. DeKruyff, H. Cantor: J. Immunol. 132, 2237 (1984)
42. D. Baltimore: In Leukemia: Recent Advances in Biology and Treatment (A.R. Liss, 1985) p. 251
43. N.A. Mitchison: Suppression of the response to murine alloantigens: Four-cell-type clusters, function-flipping and idiosyncratic responses. Progress in Allergy, to appear
44. Immunol. Rev. 95 (1987)
45. Immunol. Rev. 100 (1987)
46. M.A. Lappe: Nat. Cancer Inst. Monogr. 35, 49 (1972)
47. E.S. Golub: Cell 27, 417 (1981)
48. Z. Grossman, R.B. Herberman, S. Livnat: Neural modulation of immunity: Conditioning phenomena and the adaptability of lymphoid cells. Submitted
49. Z. Grossman, W.E. Paul: The significance of lymphokine promiscuity. In preparation
50. N.K. Jerne: In: The Neurosciences: A Study Program, ed. by T. Melnechuk and F.O. Schmitt (Rockefeller Univ. Press, 1967) p. 200
51. N.K. Jerne: Sci. Am. 229(1), 52 (1973)
52. M. Cohn: In Nucleic Acids in Immunology, ed. by O.S. Plescia and W. Brown (Springer, New York, 1968) p. 671
53. G.W. Hoffman et al.: Physica D (1987)
54. J.D. Farmer, N.H. Packard, A.S. Perelson: Physica D (1987)
55. G. Parisi: A simple model for the immune system, Proc. Nat. Acad. Sci. (USA). In press
56. G.M. Edelman: In The Mindful Brain: Cortical Organization of the Group-Selective Theory of Higher Brain Function, ed. by G.M. Edelman and V.B. Mountcastle (MIT Press, Cambridge Mass., 1978) p.51
57. S. Grossberg: Proc. Natl. Acad. Sci (USA) 77, 2338 (1980)
58. McClelland, Rumelhart and the PDP Group: Parallel Distributed Processing (MIT Press, Cambridge Mass., 1986)
59. G.M. Edelman: In Organization of the Cerebral Cortex, ed. by F.O. Shmitt et al. (MIT Press, Cambridge Mass., 1981) p. 535
60. Z. Grossman: Leuk. Res. 10, 937 (1986)
61. Z. Grossman, R.F. Levine: In Megakaryocyte Development and Function, ed. by R.F. Levine et al. (A.R. Liss, 1986) p. 51
62. Z. Grossman, R.B. Herberman: Cancer Res. 64, 2651 (1986)
63. D. Marr: Vision (Freeman, San Francisco, 1982)

Dynamical Behavior of Discrete Models of Jerne's Network

G. Weisbuch

Laboratoire de Physique Statistique de l'Ecole Normale Supérieure,
24, rue Lhomond, F-75231 Paris Cedex 05, France

The arguments exchanged during this meeting give a strong evidence that the debate about the existence and biological function of the immune network proposed by N. Jerne [1] is certainly not settled. Although the existence of idiotypic interactions is well established by experimental data, part of the debate focuses on the too simplistic dynamical properties of mathematical models supposed to take into account the most important aspects of idiotypic interactions [2]. The purpose of this intervention is to remind you that our present knowledge about networks of automata enables us to present discrete mathematical models which exhibit enough dynamical complexity to overcome the apparent paradoxes presented by the opponents to the idea of a functional immune network.

1. THE CONNECTIVITY PROBLEM

The structure of the immune network is determined by the interactions among its components, antigens, antibodies and lymphocytes. Whatever mathematical model one chooses, set of ordinary differential equations or networks of automata, the first step in an attempt to describe the immune system might be the drawing of a graph of interactions [3] . The nodes of this graph represent the cellular and chemical species of the immune system and the oriented vertices between the nodes represent the intensity of the interactions among the nodes. In the case of antibodies for instance, intensities of interactions can be represented by the affinity constants involved in law of mass action. Attempts to build such a graph have been made both from experimental data and numerical simulations. Holmberg *et al.* [4] for instance have measured affinity constants among a panel of antibodies. The numerical approach consists in random generation of chemical species defined by a sequence of random numbers, in some cases bits, in others integers. In the case of sequences of bits [3] , interaction constants among two species are measured by the maximum length of subsequences with complementary bits. De Boer's paper [5] proposes an evaluation of the interaction constants in the case of nodes defined by sets of integers.

The graph of interactions is one of the structural representations of the immune network, from which we want to derive the dynamical properties of the network. Even in the absence of a complete mathematical description, some of the network dynamics

Springer Series in Synergetics, Vol. 46 **Theories of Immune Networks**
Editor: H. Atlan and I.R. Cohen

might be predicted. The main issue concerns the connectivity of the graph. If the graph can be decomposed into non-connected subgraphs, it is clear that any perturbation of a subgraph, such as antigen presentation, leaves the other subgraphs non perturbed. On the other hand, if the graph is fully connected, one might argue that any perturbation can percolate [6] -propagate- accross the whole graph modifying tremendously the states of the nodes (concentrations in the case of continuous models, states of the automata in the case of discrete models). De Boer's contribution [5] gives an example of a system which either does not respond to a perturbation, or gives such a strong response that any memory of previous antigen presentations is lost. A way to summarize, in terms of the graph of connections, the critics of the opponents to the network theory is then to say :

- Either the connectivity of each node is small (the connectivity of a node is the number of other nodes which are connected to it), and in such case the whole network is divided in small subunits. In such a case, don't bother us with large networks and collective properties: rather use a precise description based on a set of a few ordinary differential equations for each subunit.

- Or the connectivity is so large that percolation occurs: nearly all nodes are connected together, directly, or indirectly. In this latter case, known models exhibit a behavior very rigid such as the one described by De Boer.

We shall now review some of the properties of networks of automata which avoid the two pitfalls. More precisely, our present knowledge of idiotypic interactions makes us believe that the network is indeed well connected and that nearly all of its nodes are connected together directly or not. We shall be able to specify construction rules allowing to build networks that have a rich, but robust enough behavior, allowing the system to learn, to memorize and to retrieve information in the case of secondary antigen presentation.

2. CLASSIFICATION OF THE POSSIBLE DYNAMICAL BEHAVIORS OF LARGE NETWORKS OF AUTOMATA

First of all we shall use the formalism of networks of automata since it allows to deal with large numbers of dynamical components (say 1000 in computer simulations, infinite numbers when results are exactly derived). The binary automata are of the type described by H. Atlan [7] in his contribution (boolean automata, or threshold automata), only their number is larger. We shall first recall S. Wolfram's dynamical classification of cellular automata [8] and then apply it to our case. Cellular automata considered by Wolfram are identical automata distributed on a regular lattice with connections among neighbors. Wolfram's classification is based on the properties of the attractors of the dynamics. It is applied to networks defined by the boolean function of the automata.

- The first class is composed of automata with a very *boring behavior* : all, or nearly all, configurations of automata converge in time towards a uniform configuration made only of 0's or of 1's. This is the case for AND functions, or for OR functions.

- The second class gives *organized behavior*. There are many different attractors according to the initial configurations, and their period is usually small.

- The third class is *chaotic* : the periods are very large and scale with the exponential of the number of automata of the network.

An important difference between the second and the third class is the sensitivity to initial conditions. In the organized regime, if one changes the state of only one automaton in the initial configuration, the attractor reached after the iteration differs at most in a small neighborhood of the perturbed automaton from that of the unperturbed initial configuration: perturbations remain localized. By contrast, in the chaotic regime, a small perturbation can propagate across the whole network; two configurations that differ initially in only one automaton can differ by an important fraction of the number of automata after a large number of iterations.

What we know of the biological immune system definitely indicates that any reasonable model should belong to the second class (organized behavior). Enough attractors should exist to describe various immune conditions after a number of immunizations against different antigens. Furthermore, the network should be robust enough, so that its configuration is not completely upset by new antigen presentations.

3. RANDOM NETWORKS

It is difficult to think of the immune network in terms of a network of identical cellular automata with regular connections, although some authors have already proposed such models [9]. We rather view the set of interactions among idiotypic units as random, in the sense that the connection structure is highly unpredictable from our present biological knowledge. Partial experimental investigations seem to support this idea. We shall then suppose that the connection structure and the strength of interactions among units are randomly chosen, and we shall be interested in the dynamical properties of the network which are not specific of the details of the interactions structure, but which are common to most networks built on the same principles of construction. After all, we know that immune networks of different individuals are not the same because of genetic differences and differences in the history of antigen presentations, but their behaviors present certain common features, those which are described in textbooks for instance. In other words we are looking for the generic properties of random nets such as their classification according to dynamical criteria into boring, organized or chaotic behaviors.

3.1 Ferromagnetism, an example of boring behavior

Physics provides us with the example of magnetic systems with only positive interactions, the ferromagnets [10]. In its most elementary version, the units of the system are binary spins which can be in either state ±1, and all interactions of one spin with its neighbors are positive : they favor the configurations with spins sharing the same state, either +1 or -1. It can be shown that when interactions are all positive, only two stable configurations are possible, one with all spins in state +1, and the other one with spins in state -1. In order to modify a stable configuration one must strongly perturb the system : then all spins change their state. Otherwise nothing happens. This "all or nothing behavior" is to be compared with the results of De Boer. A number of other magnetic systems which derive by some elementary change from ferromagnets exhibit this boring behavior, with only two attractors so far apart that only catastrophic perturbations can induce a change of attractor.

3.2 Dynamical structuration in random boolean nets

Random boolean nets were first proposed by S. Kauffman [11] as models for cell differentiation. The units are boolean automata, which update their binary state at each iteration step according to a boolean function of their inputs. The inputs of an automaton are the states of the automata which send their signal to it; they are defined, for instance, by the graph of interactions. The boolean function is simply defined by the table of its values according to the various possible configurations of the input set. Logical function AND, OR, NAND, NOR, XOR... are examples of boolean functions. The net is doubly random since the boolean functions of the units and their connections are randomly chosen when the network is built. S. Kauffman has shown by computer simulations that such random boolean nets exhibit either organized or chaotic behavior according to the input connectivity K (the number of inputs). Small input connectivity, up to two, gives organized behavior, while large input connectivity, bigger than three, gives chaotic behavior.

By displaying random boolean automata on a square lattice, we were able with H. Atlan *et al.* [12] , to relate organized behavior with the structuration of the whole network into subnetworks functionaly disconnected during the limit cycle. When one follows on a computer screen, (fig. 1), the evolution of the network, it can be seen that some automata remain fixed during the limit cycle. These automata are in strong

```
. 0 0 0 . . * . 0 1 * . 1 0 1 *      . 1 1 1 . . * . 0 1 * . 1 0 1 *
* 0 1 0 1 0 . . 0 1 * . 1 1 1 1      * 0 1 0 1 0 . . 1 1 * . 1 0 1 0
0 . . 0 0 0 0 . . 0 1 1 * 0 1 1      0 . . 1 1 0 1 . . 0 1 1 * 0 1 1
. . 0 0 1 0 0 * * 1 1 0 * . * .      . . 0 0 1 0 0 * * 0 1 1 * . * .
* * 0 1 * . 0 0 . 0 * * . * * .      * * 1 0 * . 0 0 . 0 * * . * * .
. 0 0 0 . . 0 1 . * 1 0 * . . *      . 0 0 0 . . 0 0 . * 1 1 * . . *
. . * 0 . * 0 0 1 . 1 1 1 . . .      . . * 0 . * 0 0 1 . 1 1 1 . . .
. 1 0 0 . 1 0 . * 0 0 * 1 . . .      . 0 0 1 . 0 0 . * 1 0 * * . . .
0 1 1 1 0 1 1 0 0 * * * 1 0 * 1      . 1 0 1 1 1 1 0 0 * * * * * * .
1 * 0 . 0 1 1 0 1 1 . 1 1 1 0 0      * * . . 0 1 1 1 1 0 . * . * * .
0 0 1 * 1 0 0 1 * . 1 0 . 1 1 1      . * * * 1 0 0 1 * . 1 1 . . * .
1 1 * * 0 1 0 . * * 0 0 * 1 1 0      . * * * 0 1 0 . * * 0 1 * * . .
* . * * 0 0 0 * . . . 0 0 1 0 .      * . * * 0 0 0 * . . . 1 0 1 1 .
* * * . * . . . * . * . 1 0 . *      * * * . * . . . * . * . 1 0 . *
. * . . . * . * * * . 1 0 1 . .      . * . . . * . * * * . 0 1 1 . .
* . . * * * * * * . . 1 1 . . .      * . . * * * * * * . . 1 1 . . .
```

Figure 1 Spatial organization of a random boolean net, H. Atlan *et al.* [12].
Two configurations of the same 16x16 boolean network, obtained for different initial conditions. The 0's and 1's are the state of oscillating automata at a given time, the * and . represent the states of automata that remain fixed during the limit cycle, respectively in states 1 and 0. It can be noted that nearly all the stable automata are members of a connected set which isolates patches of oscillating automata. Because of this barrier of stable automata, information cannot be transferred from one patch to another.

proportion of the total number of automata, and most of them are contiguous. They form a fully connected set which isolates subsets of connected oscillating nodes. Furthermore, in nearly all cases, a perturbation inside the fixed set decays very fast (perturbing one unit means inverting its state). Perturbations inside oscillating subnets might persist, but only inside the subnet. In other word, although the network is fully connected in terms of its static structure, when the attractor is reached, the net is dynamically and functionaly disconnected into independent subunits.

Such a dynamical structuration is not observed in the chaotic regime : most nodes are oscillating and their set is fully connected.

These results are also true for random connectivity, but the lack of a two dimensional representation make them more difficult to visualize. In conclusion, the organized regime is characterized by the following properties :

- small periods

- a reasonably large set of different attractors

- perturbations remain localized in only parts of the network.

These properties are prerequisites for a model of the immune system.

Important remark. The fact that we discuss about lattices of automata, and use terms such as localization, does not imply that we refer to real three dimensional space. In our description the immune system is considered as homogeneous: all immune species are supposed to be carried in a well stirred fluid. The only spatial considerations that could be introduced would concern the space of the idiotypes, but unfortunately we do not know much about its topological properties.

3.3 Threshold automata

As we said above, organized behavior is observed in random boolean nets only in the case of small connectivity. Obviously the connectivity in the immune network is larger than 2 and we have to eliminate networks of boolean automata as possible models for the immune system on these grounds. Furthermore some boolean functions, like the XOR function, seem difficult to be implemented in any simple system of interacting cells or antibodies. It is then more reasonable to use threshold automata [13] .

Let us recall that threshold automata are defined by a state transition function given by :

$$S_i = 1 \quad \text{iff} \quad \sum_j J_{ij} \; S_j > h_i$$

$$S_i = -1 \quad \text{otherwise,}$$

where S_i and S_j are the states of automata i and j, J_{ij} is the intensity of the connection from j to i, and h_i is the threshold of automaton i. In other words, automaton i takes states 1 if, and only if, the weighted sum of its inputs is larger than its threshold. The weighted sum is called local field by physicists, post-synaptic potential by neurophysiologists, sensitivity by some immunologists ...

Networks of threshold automata have been widely used as models of the nervous system under the name of neural nets and their dynamical properties have been under intense study since 1982, by means of computer simulations and analytical methods [14-17] .

The most relevant models for us are dilute nets [18,19] . In these nets, as opposed to the more classical Hopfield's model, the connectivity of each automaton is much smaller than the total number of automata of the network. Furthermore the connection structure is random. If the two conditions are fulfilled, powerful analytical methods can be used to predict the dynamical behavior of the system. We used the method of distances [18], which allows to predict the time evolution of the distance between two configurations which were initially very close (the distance between two configurations is the number of automata which are in different states). If the distance at large times remain small the behavior is organized, otherwise it is chaotic: small distances at large time imply that small perturbations of the system remain localized, a feature that we interpreted as memorizing new antigen presentation without upsetting the whole configuration.

If one considers a random dilute networks with input connectivity K and J_{ij} randomly chosen as +1 or -1, and with all thresholds equal to h, it can be shown by the distance method that if

$$h > \sqrt{KLogK}$$

the behavior of the network is organised [20] . In other words, we know at least of one model biologically reasonable in terms of the structure of interactions among units, which has a dynamical behavior both rich and robust enough to explain memory effects (the change of attractors when some units are perturbed), and robustness of previous memories when new antigens are presented. Furthermore this interesting behavior is not restricted to a particular choice of the components of the network, but is obeyed with probability close to 1 for all the nets with principles of construction as those described above.

3.4 Learning and organized behavior

All the preceding discussion implied that connections are random. Two biological reasons, when translated into network formalism, presumably modify this randomness assumption.

- since connection exists among species which recognize each other by complementarity of their epitopes, we might suppose that if J_{ij} is non-zero, J_{ji} is probably non-zero too. Even if we do not assume symmetrical connection intensities among units because of possible differences in their biological functions, many loops between two units should be present, which contradicts the hypothesis of fully random connections.

- we have biological evidence that the maturation of the immune response implies some increase of the affinity constants of the antibodies specifics for the antigen. In terms of the J_{ij}, this implies that the connection intensities of pairs of automata maintained at state 1 should increase [21]. Such a "learning" process, is reminiscent of "Hebbian learning" in neural nets : in Hebbian learning, the connections J_{ij} are changed according to :

$$J_{ij} \longrightarrow J_{ij} + (2 S_i - 1)(2S_j - 1)$$

when a configuration of activity $\{S_i\}$ is presented to the system.

Numerical simulations done by K. Kurten [22] have shown that learning in neural nets makes them change from chaotic to organized behavior. Learning in immune networks is different from learning in neural nets in that :

- the only changes that occur for immune nets are the increase of the J_{ij} when both interacting populations are in state 1.

- in formal neural nets, a pattern of activity is imposed to the whole network (in the case of Hopfield nets [14]) or to a large fraction of the network, as in the case of layered structures [23]. In both cases all the connections of the net are susceptible to change according to some prescription, Hebb's rule in the first case, or some back propagation rule in the second case. On the opposite, in immune nets, the event triggering learning is not the presentation of a whole pattern of activity, but immunization by an antigen which triggers the activation of a small part of the net. Only those cells specific of the antigen, plus idiotypically connected cells, reach concentrations large enough to allow the selection and learning process.

Although we have presently no simulation result for learning immune systems, we might infer from Kurten's results that the range of parameters (connectivity and thresholds) which result in organized behavior is increased when any type of learning occurs.

4. CONCLUSIONS

Several conclusions can be drawn from the preceding analysis.

- Sound mathematical models can be proposed to substantiate N. Jerne theory of an idiotypic immune network.

- In such models, although the graph of interaction is connected, when the attractors of the dynamics are reached, the network is functionaly disconnected into quasi-independent subunits. Perturbations remain localized in a single subunit, which results into properties of memory and robustness of the network. This localized behavior is an *a posteriori* justification of the models which take into account only a small number of interacting units, such as those presented by M. Kaufman [24] and H. Atlan [7] in this meeting.

- Pathological events such as auto-immune diseases might in some cases be triggered by the improbable occurence of a perturbation propagating from one subnet to the other. In such a case, antigen mimicry between the external antigen triggering the auto-immune response and the self-antigen would not be necessarily implied in the onset of the disease.

Acknowledgments

We thank H. Atlan, Y. Cohen, B. Derrida, F. Jacquemard, K. Kurten, A. Perelson and G. Toulouse for helpful discussions and R. de Boer for communicating his manuscript prior to publication. The author is a member of CNRS URA 1306 and acknowledges the partial support from INSERM grant 879001.

References

[1] N. K. JERNE: Towards a netwok theory of the Immune System, Ann. Immunol. (Inst. Pasteur) **125C**, 373-389, (1974).

[2] Immunol. Rev. **79**, Idiotypic networks, (1984).

[3] A.S. PERELSON: Towards a realistic model of the immune system, 377-401, in «Theoretical Immunology » part two, ed. by A.S. Perelson, Addison Wesley (1988).

[4] D.S. HOLMBERG, S. FORGREN, F. IVARS and A. COUTHINHO: Reactions among IgM antibodies derived from normal neonatal mice, Eur. J. Imunol. **14**, 435-441, (1984).

[5] R.J. DE BOER: Extensive percolation in reasonable idiotypic networks, this volume (1989).

[6] D. STAUFFER, «Introduction to percolation theory», Taylor and Francis (1985).

[7] H. ATLAN, this volume (1989).

[8] D. FARMER, T. TOFFOLI, S. WOLFRAM Eds: «Cellular Automata». Physica 10D, North-Holland, (1984).

[9] R.B. PANDEY and D. STAUFFER: Immune response via interacting three dimensional network of cellular automata, Journal de Physique (1988).

[10] C. KITTEL and H. KROEMER, «Thermal Physics», Freeman and Co. (San Francisco 1980).

[11] S. A. KAUFFMAN, J. Theor. Biol., **22**, 437-467, (1969).

[12] H. ATLAN, F. FOGELMAN-SOULIE, J. SALOMON and G. WEISBUCH: Random boolean networks, Cybernetics and Systems, **12**, 103, (1981).

[13] W.S MAC CULLOCH, W. PITTS: A logical calculus of the ideas immanent in nervous activity. Bull. Math. Biophysics, **5**, (1943), 115-133.

[14] J.J. HOPFIELD: Neural Networks and Physical Systems with Emergent Collective Computational Abilities, P.N.A.S. USA , **79**, (1982), 2554-2558.

[15] E. BIENENSTOCK, F. FOGELMAN SOULIE, G. WEISBUCH Eds: «Disordered Systems and Biological Organization», Springer Verlag, NATO ASI Series in Systems and Computer Science, n°F20, (1986).

[16] J. DENKER Ed.: «Neural Networks for Computing». Conf. Proceedings n°151: Snowbird, Utah, 1986. American Institute of Physics (1986).

[17] G. WEISBUCH, «Dynamique des systèmes complexes, une introduction aux réseaux d'automates», InterEditions (Paris 1989).

[18] B. DERRIDA and G. WEISBUCH: Evolution of overlaps between configurations in random boolean networks, J. de Physique, **47**, 1297, (1986).
[19] B. DERRIDA, E.GARDNER and A. ZIPPELIUS: An exactly solvable asymmetric neural network model, Europhysics Let., **4**, 167, (1987).
[20] B. DERRIDA: Dynamical phase transitions in non-symmetric spin glasses, J. Phys. A 20, **L721-725**, (1987).
[21] H. ATLAN, I. COHEN and G. WEISBUCH, to appear (1989).
[22] K. E. KURTEN: Training quasirandom neural netwoks, in «Chaos and Complexity», ed. R. Livi, S. Ruffo, S. Ciliberto and M. Buiatti, World Scientific (Singapore 1988).
[23] D.E. RUMELHART, J.L. MAC CLELLAND Eds: «Parallel and Distributed Processing: explorations in the Microstructure of Cognition». 2 vol., MIT Press, (1986).
[24] M. KAUFMAN, this volume (1989).

Some Reflections on Memory in Shape Space

L.A. Segel[1] *and A.S. Perelson*[2]

[1]Department of Applied Mathematics, Weizmann Institute of Science, Rehovot 76100, Israel

[2]Theoretical Division, Los Alamos National Laboratory, Los Alamos, NM87545, USA

1. Introduction

Elsewhere [1] we put forward a network model of the immune system that explicitly incorporates cross-reactivity between clones. Receptor shape and hence reactivity was described by a single real number, x, whose value might be viewed as representing the depth of the antibody combining site or the height of a complementary idiotypic "bump". Each clone was thus described by a unique value of a shape parameter x, and a rule was specified for determining the degree of reactivity of clone x with other clones y. A population dynamic model of interacting clones in a one-dimensional "shape space" was then developed that led to certain strategic insights concerning the operation of the immune system.

In a "virgin" immune system one might imagine that all clones are of equal size. This would correspond to a uniform distribution in shape space. Using our model we predicted that there is a boundary surface that divides parameter space into two regions. In one region, called *stable*, a transient fluctuation — i.e. the temporary imposition of a small change in the number of cells in various clones — would have no long term effect; the interactions between clones would return the system to its initial uniform state. In the other region, the *unstable* one, small perturbations would be amplified by the interactions between clones and a nonuniform distribution of clones in shape space would arise. We argued that an immune system that is sensitive to outside stimuli, but not unduly so, should have parameters that are somewhat (but not deeply) on the stable side of this boundary.

A recurrent feature of modeling in immunology is the demonstration that reasonable models can be constructed that contain more than one stable steady state. (Examples can be found in the volumes containing [1].) A stable steady state of low amplitude is then commonly identified with a "virgin" immune system while such a state with relatively high amplitude is identified with some sort of immune or memory state. The virgin state might then correspond to a simple balance between cell supply from the bone marrow and natural cell death. The immune or memory state would be a *dynamic* state in which the cell population was maintained at an elevated level due to interactions among various clones, e.g. idiotypic anti-idiotypic pairs or more complex network interactions.

This maintainance of elevated clonal populations via dynamic interactions contrasts with the classical *static* view of memory held by many experimental immunologists who have found evidence for the existence of "memory cells". Memory cells are cells of relatively long lifetime, or perhaps even immortal cells, that are somehow brought into being when a virgin clone is stimulated by antigen. Here we wish to employ our model [1] to explore the matter of how memory cells fare in a network environment. Assuming that memory cells are identical to virgin cells in all aspects except that they live longer, we are interested in the question of whether network interactions will allow memory clones to be maintained at elevated levels or will the network try to self regulate in such a way that memory clones will be reduced in population size so as to approach a common background level.

2. Formulation

One can conceive of the immune system as a collection of molecular populations each of which is characterized by a generalized shape (including charge distributions). We consider for simplicity a one-dimensional shape space, and we assume that the smaller is $x-(-y) \equiv x+y$, the tighter the fit there is of shape x with shape y. Thus $y=-x$ is exactly complementary to x. See Fig. 1. Our fundamental unknown function will be $b(x,t)$, the number of lymphocytes with receptors of shape x that constitute the immune system at time t. Antibody can be regarded as present if it is assumed that there always is a constant number of solution-phase antibodies per lymphocyte.

The cell population is divided into two classes, stimulated (probability α) and unstimulated (probability $1-\alpha$). Unstimulated cells die at rate d, while the bone marrow supplies new cells at a constant rate m. Stimulated x-cells reproduce at a rate r that is a function not only of the population level $b(x,t)$ of such cells but also, to account for nonspecific factors, of the average lymphocyte population size $B(t)$. We can represent antigens of shape x at time t by a source function In the $b(x,t)$ equation or with a separate population dynamic equation. However, antigen is neglected in the calculations to be described here. With all this, the basic equation of the model is

$$\frac{\partial b(x,t)}{\partial t}=m-db(x,t)\{1-\alpha[b(x,t)]\}+r[B(t),b(x,t)]\cdot\alpha[b(x,t)]\cdot b(x,t). \tag{1}$$

If shape ranges between $-L$ and L, then the average population size B is given by

$$B(t)=\frac{1}{2L}\int_{-L}^{L} b(x,t)\,dx. \tag{2}$$

To specify the function α, which depends implicitly on the distribution of clones in shape space $b(x,t)$, we assume that whether or not a cell is switched into an active state depends on a competition between activating and suppressing influences, A and S. Although the mechanisms by which suppressive influences act are not fully elucidated, here we have chosen to model activation and suppression as operating via two distinct receptor systems. For any fixed shape x, let $a(x,y)b(y,t)\,dy$ be the fraction of an x-cell's activating receptors that are

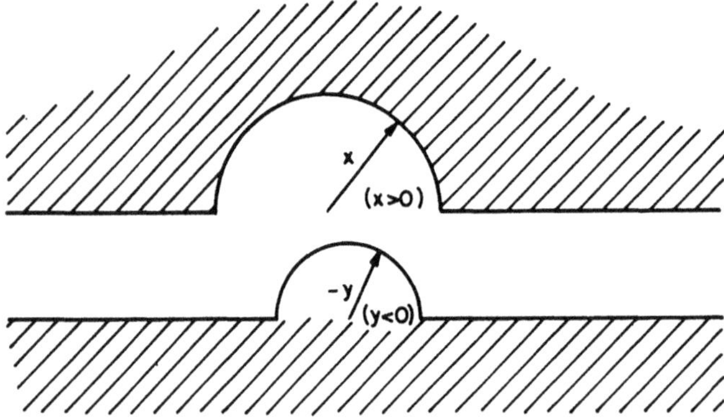

Fig. 1. A possible realization of one-dimensional shape space. Domains of molecules that are important for binding are idealized as hemispherical indentations or protuberances. The coordinate can be taken as the radius of an indentation and the negative of the radius of a protuberance.

bound by cells whose shape variable ranges between y and $y + dy$, where dy is a small number. Then

$$A(b) = \begin{array}{l}\text{fraction of } x\text{-cells'}\\ \text{activating receptors}\\ \text{bound when cell}\\ \text{distribution in shape space}\\ \text{of cells is given by } b(y,t)\end{array} = \int_{-L}^{L} a(x,y)b(y,t)\,dy. \tag{3}$$

In essence $a(x, y)$ is the association constant for y binding to activating receptors on x-cells (assuming that a small fraction of receptors is bound).

Analogously, for the suppressing influence

$$S(b) = \int_{-L}^{L} s(x,y)b(y,t)\,dy. \tag{4}$$

The probability of stimulation α is some function F of the amount of binding to the two classes of receptors:

$$\alpha = F(A, S). \tag{5a}$$

The function α increases (decreases) if more activator (suppressor) receptors are bound:

$$\partial F/\partial A > 0, \qquad \partial F/\partial S < 0. \tag{5b, c}$$

A simple possibility for F is the Michalean expression

$$F(A, S) = \frac{A}{(p + qS) + A}. \tag{6}$$

To embody our assumption that the fit of shape y to shape x decreases from a maximum when $y = -x$ (perfectly complementary shape), we assume Gaussian functions of $x - (-y)$ for the activation and suppression association factors A and S, with standard deviations σ_a and σ_s respectively. An important parameter is the ratio

$$\theta \equiv \sigma_s^2/\sigma_a^2. \tag{7}$$

For the proliferation rate r of activated cells we generally take

$$r(B, b) = r_0 e^{-\eta B^n} e^{-\lambda b} \qquad (r_0 = \text{constant}). \tag{8}$$

The calculations to be presented here were carried out in the absence of a nonspecific effect on growth rate ($\eta = 0$).

3. Analysis

Numerical calculations are carried for a discrete version of (1)

$$db_i(t)/dt = m - db_i(t)[1 - \alpha] + r[B(t), b_i(t)] \cdot \alpha \cdot b_i(t). \tag{9}$$

In discretizing the equations that supplement (9), integrals are replaced by sums. For example the activating contribution to the fraction $\alpha(A, S)$ of stimulated cells is [replacing (3)]

$$A = \sum_{j=-N}^{N} a_{ij} b_j. \tag{10}$$

The interval $-L \leq x \leq L$ is broken into subintervals of width Δ,

$$\Delta = L/N, \tag{11}$$

with b_j assigned values at the $2N + 1$ interval endpoints. To minimize "edge effects" near $j = -N$ and $j = N$ we make the periodicity assumption $b(i) = b(i + 2N)$.

The actual immune system consists of a large number of clones, so that in one sense the continuous formulation (1) can be regarded as a relatively tractable approximation to the "real" equations (9). Perhaps better, however, is to regard (1) as an approximation to a system with a number M of clones, where $M \gg 2N$. Then (9) approximates (1), i.e. (9) is an approximate system wherein each "clone" in fact represents many clones of similar shapes.

The governing system of equations possesses a uniform positive steady state $\bar{b}$ [1]. In the cases we shall consider below, when $b = \bar{b}$ then $r \approx r_0$, $F(\bar{b}, \bar{b}) \approx q^{-1}$. Thus

$$\bar{b} \approx m/[d - q^{-1}(d + r_0)]. \tag{12}$$

Further, as detailed in the caption to Fig. 2, $q = 50$, and thus only 2% of cells are activated in the uniform steady state.

The steady state is stable to small-amplitude perturbations providing $r_0 > r_0^{crit}$, where r_0^{crit} is a certain function of the remaining parameters. When r_0 is just barely greater than r_0^{crit} then calculations provide the wavenumber k_c (and the wavelength $2\pi/k_c$) of the perturbation during its initial period of exponential growth.

4. Results

We have argued [1] that an immune system should be stable but controllable. We thus suggested that parameters should have evolved so that an unstimulated immune system is stable to small transient perturbations but not overly stable. That is, the parameters should be set not too far into the stable domain of parameter space:

$$r_0 < r_0^{crit} \quad \text{but not} \quad r_0 \ll r_0^{crit}.$$

Given this, upon beginning the present investigation we felt that if a clone of memory cells were indeed to persist in a relatively expanded state then its parameters would have to be shifted into the unstable parameter domain. In other words, our intuition was that a memory clone might be an unstable "island" in a stable "sea". If this were not the case, we conjectured that the stabilizing influences of network interactions would return the system to a very nearly uniform state and memory would be lost. Let us examine how this view of unstable memory islands bears up in the face of computer simulations of our model.

Memory cells are long-lived cells. Here we model them as being identical to virgin cells except that we assign them a lower death rate, d. If one examimes the theory presented in [1], it is easy to see that lowering the death rate of a clone increases the steady state population level of the clone (cf. 12) and moves the parameters characterizing that clone toward or into the unstable parameter regime. In what follows we chose a single clone to be a memory clone and assign it a death rate that places the clone near or over the instability boundary.

Fig. 2 depicts a situation of an immune system that is set only slightly below the threshold of instability. To be precise the actual value of the basic birth rate parameter r_0 is six percent below the critical value for instability. Virgin clones have a death rate $d = 1.3$. One clone $i = -14$, has been chosen as a memory clone, and has been assigned a lower death rate,

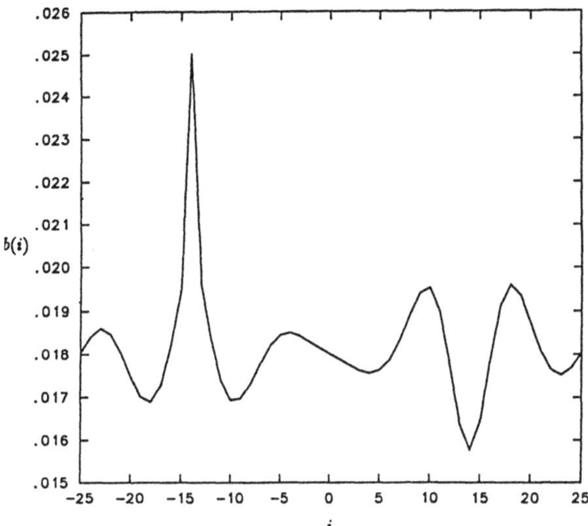

Fig. 2. At $i = -14$, r_0 is about 2% above its critical value $[r_0^{crit} - r_0)/r_0^{crit} = 0.022]$. For all other i, r_0 is about 6% below critical $[(r_0^{crit} - r_0)/r_0^{crit} = 0.056]$. Also $m =$ 0.01, $\lambda = 0.5$, $r_0 = 37$, $\sigma_a = 0.1$, $\sigma_s = 0.033333$, $q = 50$, $\Delta = 0.028$, $N = 25$, $p = 0.0001$, $d_i = 1.3$ for $i \neq -14$, $d_{-14} = 1.2$.

$d = 1.2$, that puts this clone about 2% above the critical value for instability. The system is one in which inhibition is relatively short range—the parameter θ of (7) has the value 1/9. What we see is that indeed the unstable clone has expanded significantly, but the disturbance has propagated extensively throughout the entire system. In particular, there is significant suppression in the neighborhood of the complementary clone $i = 14$. In the absence of a memory clone this system settles down to a uniform steady state wherein $b_i = 0.018$ for $-N \leq i \leq N$.

In Fig. 3 we have set the *background clones* ($i \neq -14$) into a considerably more stable state, almost 40% below critical, by increasing their death rate to $d = 2$. The uniform steady state now occurs with $b_i = 0.008$. The disturbance stemming from the unstable clone $i = -14$ is much more limited, while the amplitude of this clone is even slightly enhanced.

We now increase the death rate of the memory clone $i = -14$ to $d = 1.3$ so that it too is stable; indeed, like the background in Fig. 2, its birthrate is now six percent below critical, whereas clones with $i \neq -14$ are still approximately 40% below critical. As shown in Fig. 4, the memory clone is still strongly amplified!

To study the effects of our discretization the total number of clones is doubled in Fig. 5, with the memory clone now being $i = -39$. The sharp peak is again confined to a single clone, and the disturbance to the complementary clones is somewhat damped.

In Fig. 6, conditions are the same as in Fig. 3 except that the ranges of activation and inhibition are reversed. Now $\theta = 9$. Only one major change is seen—the complementary clones centered around $i = 14$ are now enhanced, to almost the same extent as the suppression evidenced in Fig. 3. (Ref [3] provides a detailed discussion of the influences of relatively short-range and relatively long-range activation.)

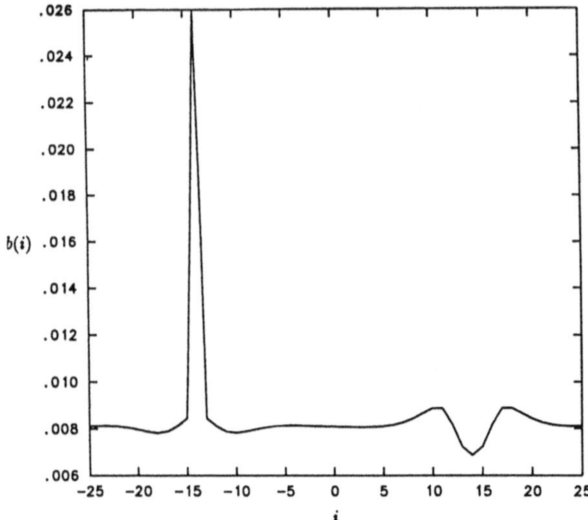

Fig. 3. As in Fig. 2, except that the "background" ($i \neq -14$) is now much more stable $[(r_0^{crit} - r_0)/r_0^{crit} = 0.38]$. *Also* $d_i = 2$ *for* $i \neq -14$.

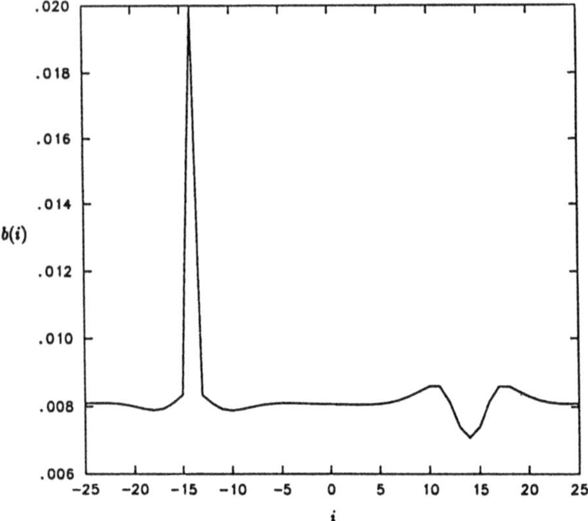

Fig. 4. As in Fig. 2 except that at $i = -14$, r_0 *is now about 6% below its critical value. Also* $d_{-14} = 1.3$.

Conclusions

Examination of our results indicates that despite our initial misgivings, expanded memory clones *can* exist as islands in a sea of stable clones. Moreover, systems with a rather stable background behave in a rather simple fashion.

If the parameters of a single "memory" clone are set to less stable values by decreasing the death rate of the cells in the clone, then that clone will settle at an increased population level

Fig. 5. As in Fig. 3, except $N = 50$.

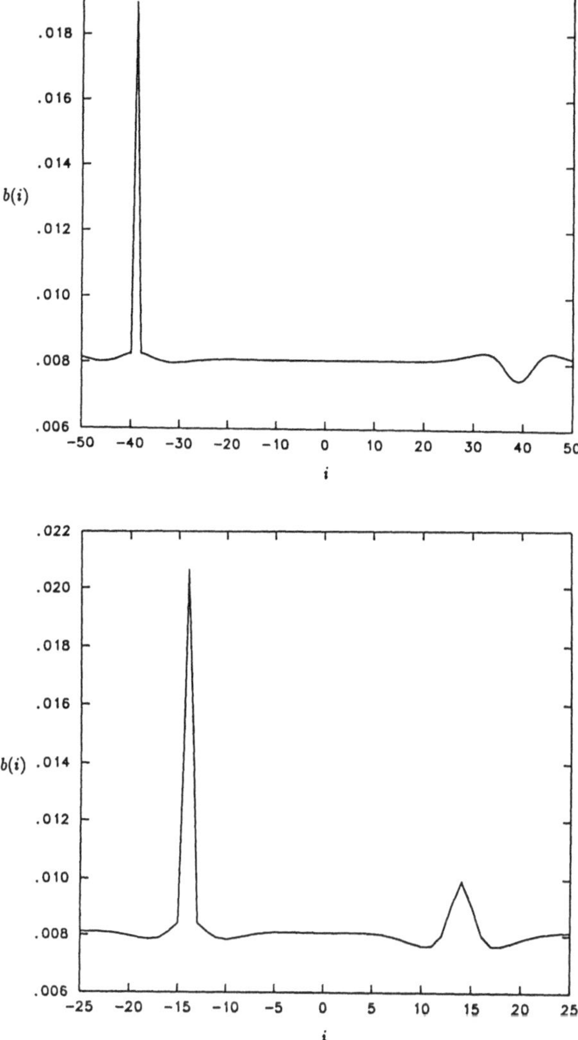

Fig. 6. As in Fig. 3, except σ_a *and* σ_s*, the ranges of activation and inhibition, are reversed* $[\sigma_a = 0.033333,\ \sigma_s = 0.1]$.

which is approximately that of a system composed of a single clone with those parameters. For the present cases, with the parameters as given, we see by employing (12) that the unstable clones with $d = 1.2$ and $d = 1.3$, respectively, would have amplitudes of approximately 0.019 and 0.023 if these clones were completely unlinked to other clones. Our simulations show that the memory clone attains a steady state of approximately this size. Further, because of our choice of parameters p and q in (6) almost all of the cells in these clones are resting ($F \ll 1$), as is typical of memory cells.

Although Figs. 2–6 show that the network hardly affects the steady state level of memory, we find that the elevated population level of a memory clone does effect the rest of the network to some extent, particularly in the neighborhood of complementary, or anti-idiotypic clones.

This effect is rather significant if the system is set only slightly below the stability threshold (Fig. 1). Making the system somewhat more stable confines the network effect of memory cells largely to excitation of their complement (Fig. 3).

Although nonlinear dynamical interactions can give rise to systems with multiple steady states, e.g. virgin, immune, and suppressed, our results show that these types of dynamic interactions are not required for memory generation if the lifetime of a cell can be changed due to interaction with antigen. The analysis presented here does not argue for or against static vs dynamic memory. We have only shown that the classical notion of a memory cell is compatible with our simple network model of the immune system. It remains to be seen whether our conclusions remain valid when our model is generalized.

Acknowledgement

This work was performed under the auspices of the US Dept of Energy and was supported by Grant 3777 of the US-Israel Binational Foundation.

References

[1] Segel, L.A. and Perelson, A.S. (1988). Computations in Shape Space. A New Approach to Immune Network Theory. In *Theoretical Immunology, Part Two*, A.S. Perelson, ed. Addison-Wesley, Redwood City, CA, pp. 321–343.

[2] Segel, L.A. and Perelson, A.S. (1989). Shape space analysis of immune networks. In *Theoretical Models for Cell to Cell Signalling* (A. Goldbeter, ed). N.Y.: Academic Press, in press.

[3] Segel, L.A. and Perelson, A.S. (1989). A paradoxical instability caused by relatively short range inhibition. *SIAM J. Appl. Math*, in press.

Part III

Regulation of the Immune Response: A Discrete Mapping Approach

M. Kaufman

Université Libre de Bruxelles, Faculté des Sciences,
Service de Chimie Physique, Campus Plaine, C.P. 231,
Bd. du Triomphe, B-1050 Bruxelles, Belgium

1. Introduction

Experimental studies have shown that prior exposure of a host to antigen modifies its subsequent antibody-response to the same antigen. In particular, low or high primary antigen doses can lead to reduced secondary responses, while medium doses increase the sensitivity to a next challenge /1,2/. In more general terms, a first encounter of a host with antigen can induce either a state of immunological memory or a state of immunological paralysis, depending on the nature or form of the antigen, on the dose and the way in which it is administered, and on the state of the host. Both these memory and paralysis phenomena reflect the behavior of the cells that interact in the course of an immune response, namely the B and T lymphocytes. B lymphocytes repond to antigen by proliferating and differentiating to plasma cells, the terminal antibody producing cells. T lymphocytes function as regulatory cells and can be classified into T helper cells (T_H) which cooperate with the B cells to bring about an effective response, and T suppressor cells (T_S) which exert inhibitory effects thereby contributing to the decay in time of the response. The outcome of an immune response should depend crucially on the balance between these two competing effects.

Our purpose has been to investigate to what extent one may account for memory and paralysis on the basis of a simple cellular network. We have considered a limited number of cell types, treating as a single entity all the cells belonging to the same category, and a small set of interactions which we believe to be especially relevant in generating a response to a given antigen /3,4,5/.

Our model relies on the following facts which are summarized in fig.1 :

1. Early immature B lymphocytes are steadily produced by the bone-marrow. At this stage the B lymphocytes are highly sensitive to inactivation and encounter with the antigen corresponding to their specificity has a negative effect /6/. The cells which escape this inactivation reach the "virgin" B_1 stage.

2. Antigen triggers the B_1 cells having receptors for it to proliferate and to differentiate to monospecific antibody-producing cells. These different steps require B-T_H cooperation : T_H cells that recognize the antigen associated with the B cells or the idiotype on their

Springer Series in Synergetics, Vol. 46 **Theories of Immune Networks**
Editor: H. Atlan and I.R. Cohen

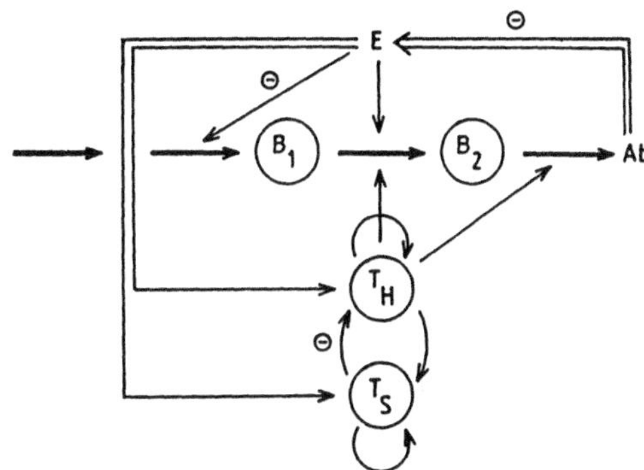

Figure 1 : The circles indicate the cellular species. The T_H and T_S compartments are assumed to involve idiotypic and anti-idiotypic cells. The dubbel arrows designate the differentiation process of the B cells and the antibody-antigen reaction. Simple arrows correspond to regulatory interactions which are positive unless otherwise specified.

immunoglobulin-surface receptors, are activated to secrete lymphokines which promote B cell growth and differentiation. We limit the description arbitrarily to two developmental stages of the B cells : virgin B_1 cells that are beyond the stage of susceptibility to tolerance induction, and activated B_2 cells that will proceed to terminal maturation provided they receive the necessary T-cell help.

3. The antibody that is produced neutralizes the antigen thus impairing its further inducing effect.

4. The T helper cells that collaborate with the B cells result from the stimulation of precursor cells that were selected by the antigen or by receptor-receptor interactions, and similarly for the T suppressor cells that interact with the T_H cells expressing complementary receptors /7/. Suppression is here considered to act mainly on the development of the T_H population.

5. T-T interactions occur within each T compartment /8/. We consider that they have a stimulatory influence /9/ and describe these positive regulations simply by an autocatalytic loop on the T_H and on the T_S cells. A more elaborate version of the model involving explicitly idiotypic and anti-idiotypic helper and suppressor lymphocytes, with cross-stimulation between complementary partners, is presented elsewhere /10/.

6. The cellular interactions are governed by antigen recognition or by receptor-receptor recognition but the stimulatory or inhibitory character of the transmitted signals is determined by the nature of the interacting cells.

Our approach is a global one and we do not intend to account for every aspect of a humoral immune response. Rather our main goal is to bring out some significant feature of the network that might explain how the magnitude of an antigenic stimulus determines the kind of response induced.

2. The Helper-Suppressor Circuit : Logical Approach

Let us first focus on the T cell circuit of the network. To study the qualitative properties of this regulatory core we have applied a generalized logical method which involves multivalued logical variables describing the concentrations of the cellular components, and logical parameters characterizing the strength of the various interactions. A detailed presentation of the method as well as its relation with a description in terms of piecewise linear differential equations can be found in refs. by Snoussi /11/ and Thomas /12/. Let us only mention that time remains a continuous variable and that the dynamical behavior, starting from a set of initial concentrations, can be followed on an asynchronous iteration graph. Whenever more than one variable has an order to switch its value, a "race" is engaged between these variables and a given state can have different successors according to the result of this race, i.e. according to the choice of the "on" and "off" time delays for each variable /13/. The rationale behind the use of a multileveled logic, on the other hand, relies on the fact that when a given element acts at several points of the interaction graph, each interaction may require a distinct threshold, and similarly each connection may correspond to a different weight /11,12/.

Considering the core of our network, we associate to the real variables x_1 and x_2 describing the concentrations of T_H and T_S cells, the discrete variables :

$$\mathbf{x_1} = d_1(x_1) \text{ and } \mathbf{x_2} = d_2(x_2)$$

where subscript 1 or 2 for the discretization operator d indicates that the discretization scales for x_1 and x_2 correspond to different sets of thresholds, T_{11} and T_{21} for x_1, T_{12} and T_{22} for x_2. $\mathbf{x_1}$ and $\mathbf{x_2}$ thus take the value 0, 1 or 2 depending on the location of x_1 and x_2 with respect to their threshold values. These multileveled variables describe the state of the system, however, for a given interaction the important point is wether the level is below or above the threshold corresponding to that interaction. This defines a corresponding set of binary variables :

$$\mathbf{x}_j^{(i)} \in \{0,1\}, \quad \mathbf{x}_j^{(i)} = 1 \text{ if } \mathbf{x}_j \geq i, \text{ and}$$

$$\bar{\mathbf{x}}_j^{(i)} = 1- \mathbf{x}_j^{(i)} \text{ with } i,\ j = 1,2.$$

After ordering the thresholds for each variable by magnitude, the logical relations for the rate of production of x_1 and x_2 are derived from the interaction graph, together with information on how the various interactions are connected to one another /11/.

Let us consider the case where $T_{11} > T_{21}$ and $T_{22} < T_{12}$. The following relations describe our system :

$$X_1 = d_1 \, (k_1 \, \bar{x}_2{}^{(2)} + k'_1 \, x_1{}^{(2)}) \qquad (2.1)$$

$$X_2 = d_2 \, (k_0 + k_2 \, x_1{}^{(1)} + k'_2 \, x_2{}^{(1)}) \qquad (2.2)$$

where the plus sign refers to the usual arithmetic sum. The parameters k,k' are positive real numbers characterizing the "amplitude" of each connection. The antigen concentration is here included in parameters k_0 and k_1.

The production rates are thus determined by the discretized value of pseudo-boolean expressions and the corresponding thruth table is :

x_1	x_2	X_1	X_2
0	0	K_1	K_0
0	1	K_1	K'_2
0	2	0	K'_2
1	0	K_1	K_2
1	1	K_1	K''_2
1	2	0	K''_2
2	0	K''_1	K_2
2	1	K''_1	K''_2
2	2	K'_1	K''_2

The stable and transient states are determined by the values of the logical parameters :

$K_1=d_1(k_1)$, $K'_1=d_1(k'_1)$, $K''_1=d_1(k_1+k'_1)$, $K_0=d_2(k_0)$,

$K_2=d_2(k_0+k_2)$, $K'_2=d_2(k_0+k'_2)$ and $K''_2=d_2(k_0+k_2+k'_2)$.

In fig.2 we have considered that the interactions not depending on the antigen concentration are strong, i.e.:

$d_1(k'_1)=d_2(k_2)=d_2(k'_2)=2$, and as a consequence

$K'_1=K''_1=K_2=K'_2=K''_2=2$.

Figure 2(a) corresponds to the absence of antigen : $K_0=K_1=0$. There are three attractors characterized respectively by the absence of T_H and T_S cells, the presence of T_S cells only, and a balance of both T_H and T_S cells. As a function of its past antigenic history, the system will be in one or the other of these stable regime states. State (00) may be interpreted as a virgin state (V) found in animals that have not yet been exposed to the antigen considered; state (02) as a suppressed state (S) found in animals that have been brought into a state of immune paralysis; and state (22) as a memory state (M) where helper T cells, needed for the triggering and differentiation of the B cells, are already present as a result of a previous encounter with the antigen.

In the presence of antigen there remain two attractors : the suppressed and memory states. The dynamical evolution starting from the naive state is determined by the values of K_0 and K_1 which reflect the activation rates of T_S and T_H cells as a function of the antigen level. For $K_1 < 2$, fig. 2(b), only the suppressed state S can be reached, whereas for $K_1=2$, fig. 2(c), both states become accessible and the final state will depend on the choice of the time delays. Let us mention here that a cruder modelling in terms of binary asynchronous automata does not allow to distinguish between the situations at low and high antigen levels /3/.

The dynamical pathways upon antigen injection and the resulting situations after disappearance of the antigen by slow spontaneous decay or neutralization are shown in fig.3. Similar results are obtained with other thresholds inequalities, given appropriate choices for the logical parameter values.

While being essentially qualitative this prior discrete treatment allows to determine the key elements responsible for a given behavior. It predicts parameter ranges to introduce in a differential description, using steep sigmoidal functions to represent the various cell interactions, in order to observe a similar behavior. The agreement between the two approaches can easily be visualized by comparing the state tables of the logical descrip-

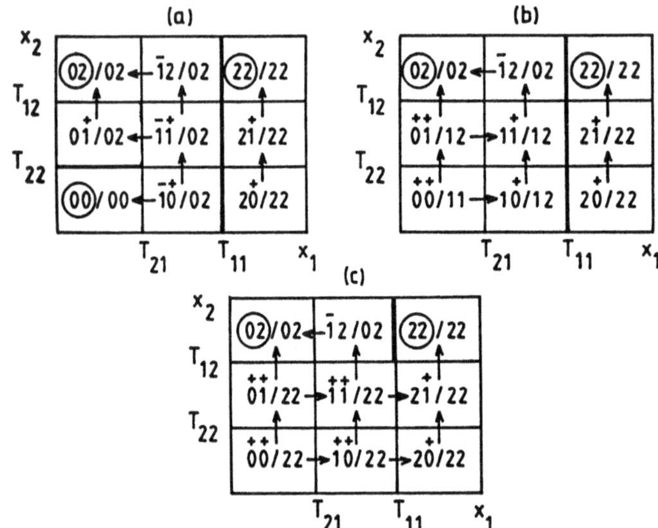

Figure 2 : State tables for different antigen levels. (a) : no antigen, $K_0=K_1=0$; (b) : low antigen level, $K_0=K_1=1$; (c) : high antigen level, $K_0=K_1=2$.

No antigen | low antigen level | high antigen level

$K_0=K_1=0$ | $K_0=K_1=1$ | $K_0=K_1=2$

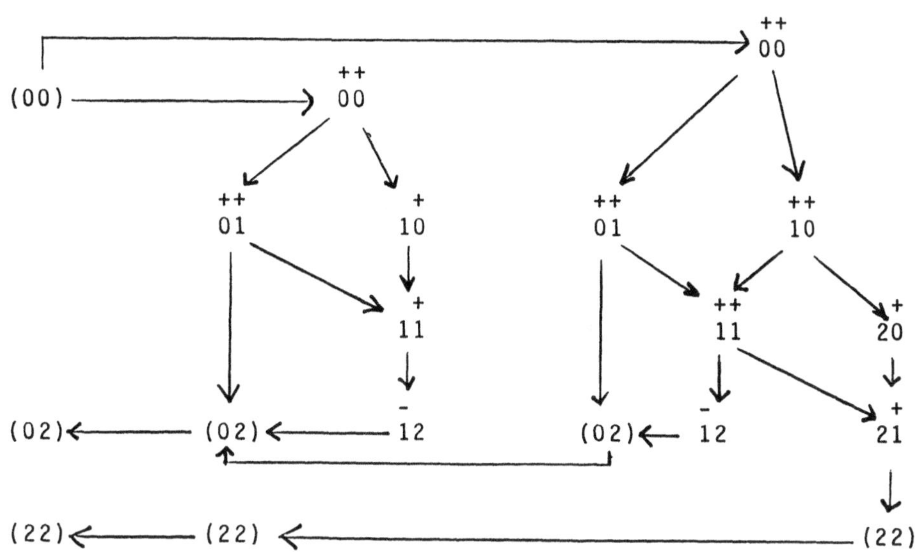

Figure 3 : State transition graphs obtained by a combination of the state tables 2(a), 2(b) and 2(c).

tion with the nullclines of the differential description (see e.g. refs./5/ and /12/), and will be discussed for the present case in section 3.

3. The Complete Network

A boolean analysis of a model quite similar to the one in fig. 1 has been given in a previous publication / 3/. Here we present a continuous analysis of the complete network and show on a few examples how the properties of the regulatory T cell core determine the outcome of an immunization as a function of the magnitude of the antigenic challenge and prior exposure to the antigen.

We adopt a global description in terms of the following set of differential equations :

$$dx_1/dt = k_1 F_{16}\, F_{\overline{12}} + k'_1\, F_{11} - d_1 x_1 \qquad (3.1)$$

$$dx_2/dt = k_0 F_{26} + k_2\, F_{21} + k'_2\, F_{22} - d_2 x_2 \qquad (3.2)$$

$$dx_3/dt = k_3 F_{\overline{06}} - k_4 x_3 x_1\, F_{38} - d_3 x_3 \qquad (3.3)$$

$$dx_4/dt = k_4x_3x_1\, F_{38} + (k'_4 - k_5)x_4x_1 - d_4x_4 \qquad (3.4)$$

$$dx_5/dt = k_5x_4x_1 - k_6x_5x_6 - d_5x_5 \qquad (3.5)$$

$$dx_6/dt = - k_6x_5x_6 - d_6x_6 \qquad (3.6)$$

x_1 to x_6 correspond, respectively, to the concentrations of the species T_H, T_S, B_1, B_2, free antibody (Ab) and antigen (E). Constant levels of precursor cells are included in the corresponding kinetic constants. The detailed mechanisms of antigen presentation and lymphocyte interactions are not described. These stimulatory or inhibitory processes are represented globally by sigmoidal functions of the Hill type:

$$F_{ij} = \frac{x_j^n}{T_{ij}^n + x_j^n}, \qquad F_{ij}^- = 1 - F_{ij}$$

with T_{ij} a threshold parameter for the regulation of component i by j, and the Hill number n characterizing the steepness of the functions. F_{06}^- represents the blocking of the influx of virgin B_1 cells by the antigen.

Stimulated B_2 cells are assumed to bear both an antigenic determinant resulting from the processing of the antigen, and an idiotypic determinant. They focus the interactions with the T_H cells and we treat the concentration of T lymphokines acting on the B cells as being proportional to the concentration of the T_H cells that produce them. The antibody binds irreversibly to the antigen thus forming an inactive complex.

Equations (3.1) and (3.2) are the differential counterpart of the logical relations (1). For very large n and parameter values corresponding to the logical analysis /11/, they present exactly the behavior predicted in section 2. The basic results remain however valid for a large range of values of n and we consider here smooth Hill functions, with n=2 for all the sigmoidal interactions.

The dynamic responses upon antigenic stimulation are obtained by numerical integration of (3.1) to (3.6), together with initial conditions, and are illustrated below for the following parameter choice :

$k_1=k_0=.15 \quad k'_1=.5 \quad k_2=.2 \quad k'_2=.25$

$k_3=1. \quad k_4=.1 \quad k'_4=k_5=.4 \quad k_6=.2$

$d_1=d_2=d_4=d_5=.2 \quad d_3=.1 \quad d_6=.05$

The threshold parameters T_{ij} are equal to one, except $T_{06}=.1$ and $T_{21}=T_{22}=.5$

Low dose paralysis.

Figures 4 and 5 illustrate the phenomenon of low dose paralysis. The initial state corresponds to a virgin system characterized by a steady concentration of virgin B_1 cells and the absence of all other species. The system is subjected at time t=0 to a low sub-immunogenic antigen dose, $x_6^*=1$, and at time t=50 to an antigen dose that normally evokes a response in an untreated system.

As predicted by the logical analysis, the first injection leads, without the appearance of antibody, to the induction of the suppressed state S, in which the system remains after slow degradation of the antigen (figs. 4(a), 5(a)). The response to a second optimal challenge is shown in dotted lines in fig. 4(b)

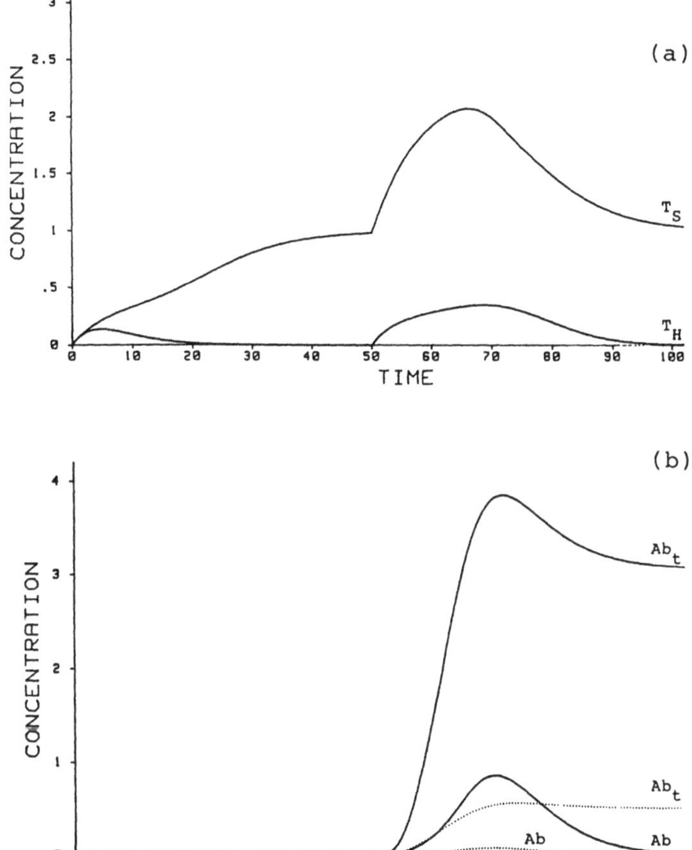

Figure 4 : (a) Evolution of T_H and T_S cells. A low dose induces the S state, in which the system remains after a second optimal injection. (b) Corresponding evolution of free antibody (Ab) and total antibody (Ab_t) . Ab_t reaches a plateau level where all the antibody is under the form of an inactive complex. (---) suppressed, (——) primary response.

for x_6*=50 and in fig. 5(b) for x_6*=250, together with the primary response that is obtained for the same dose in a virgin system (full lines). In both cases the maximum antibody production is about 10 to 20% of the corresponding normal response. These low responses are due to the inhibition of the development of the helper population needed for the differentiation of antibody-producing cells.

Recovery from low dose paralysis is not spontaneous. It may however result from the secondary challenge, depending on the strength of the stimulus. As illustrated in fig. 5(a), suppression is lifted by a strong secondary dose. The system then evolves toward the memory state M from where low dose paralysis can no longer be achieved. For lower secondary doses, as in fig. 4(a), the system settles back in the suppressed state S and will again respond poorly to a next challenge.

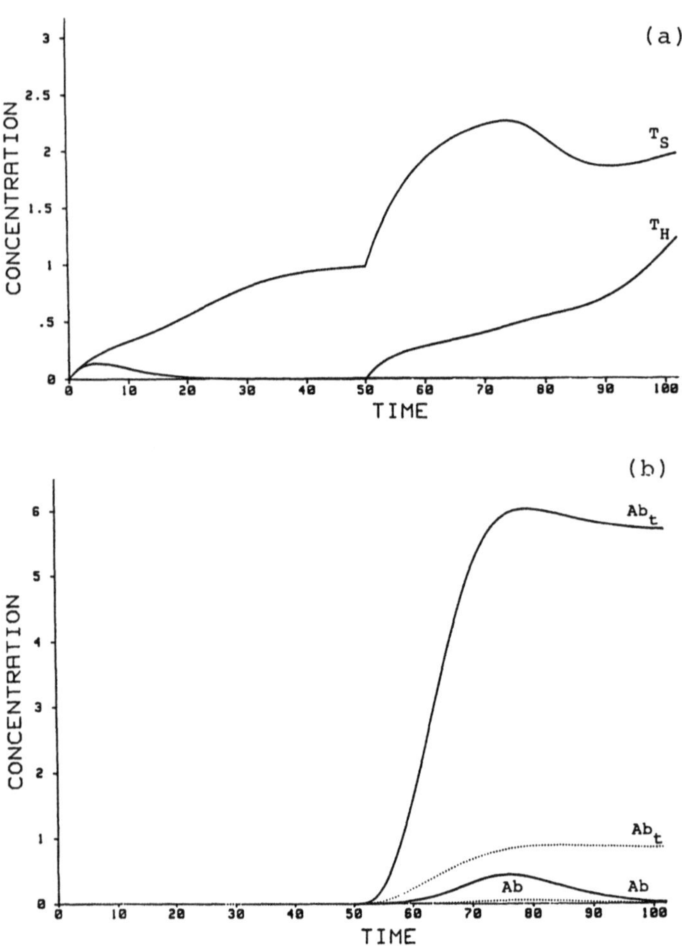

Figure 5 : (a) The first, low stimulation, induces the S state but suppression is now lifted after the second injection. (b) Suppression results again in a low antibody-response.

The fact that in our continuous analysis suppression can be lifted, contrary to what is the case in the discrete description, is related to the low value chosen for n.

Memory : enhanced secondary responses.

Figures 6(a) and (b) give the result of two identical antigenic stimulations corresponding to x_6*=250, at a time interval Δ t=50. The first stimulation leads to an effective primary response together with the establishment of the memory state M for the T cells.

Antibody is produced more rapidly and in larger amounts during the secondary encounter with the antigen. This enhanced immunity

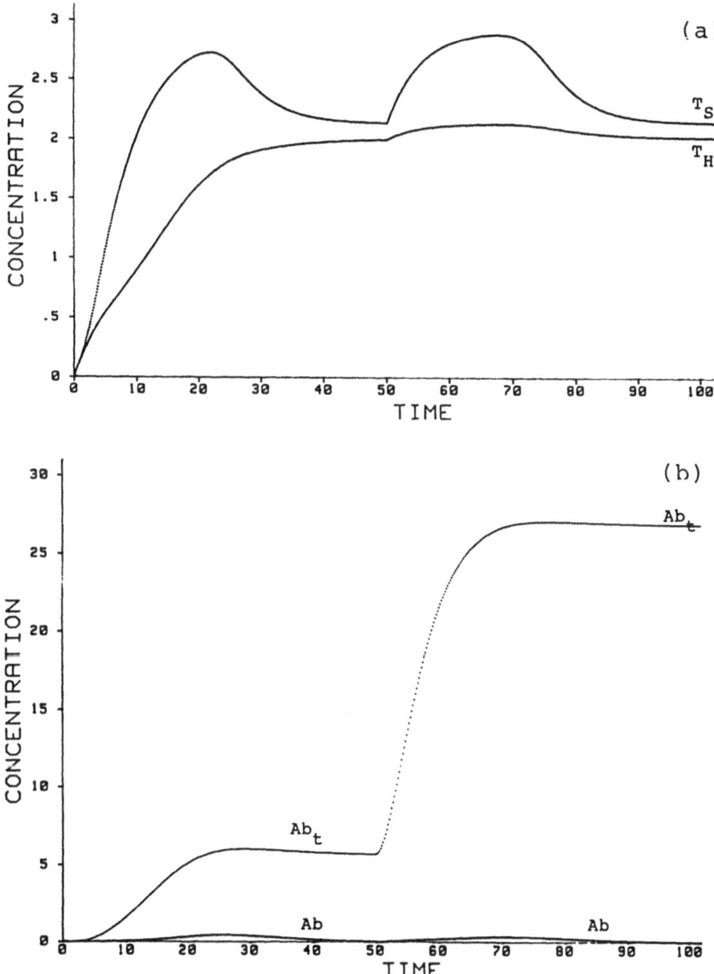

Figure 6 : (a) Time evolution of T_H and T_S concentrations upon two successive challenges. The M state is established. (b) Corresponding antibody-response : T memory leads to an increased sensitivity.

is here due solely to the T_H memory since differentiation of B cells to memory cells has not been taken into account.

High dose paralysis.

The magnitude of the primary response increases with the strength of the stimulus and reaches a plateau value for very high antigen doses. At these high antigen concentrations, the immature B cells are rapidly inactivated but the virgin B_1 cells that are already present continue to differentiate in the presence of T help and antigen. The antibody that is formed binds rapidly to the antigen and all the antibody is found under the form of antibody-antigen complex. Decreasing primary responses for high antigen doses could possibly be observed by considering a direct action of the T_S cells on the B cells or/and T_S inhibition of the autocatalytic development of the T_H cells. These points are presently under investigation.

High dose paralysis is however observed, for our model, after pretreatment of the system. Starting from the virgin state, a constant antigen level, $x_6=100$, is maintained in the system during a time interval $\Delta t=40$, followed by a rest period of 20 time units. Then the system is challenged with an optimal dose $x_6^*=250$. The result of pretreatment is shown in fig. 7. The memory state is achieved, nevertheless no antibody is produced in response to the optimal stimulation.

The same pretreatment with $x_6=1$ and $x_6=.5$ leads after challenging with $x_6^*=250$, respectively, to a response about 3 times larger than the corresponding primary response, and 10 times lower than the primary one. As a function of the dose used for pretreatment there are tnus three successive regions corresponding to low dose paralysis, immunity and high dose paralysis.

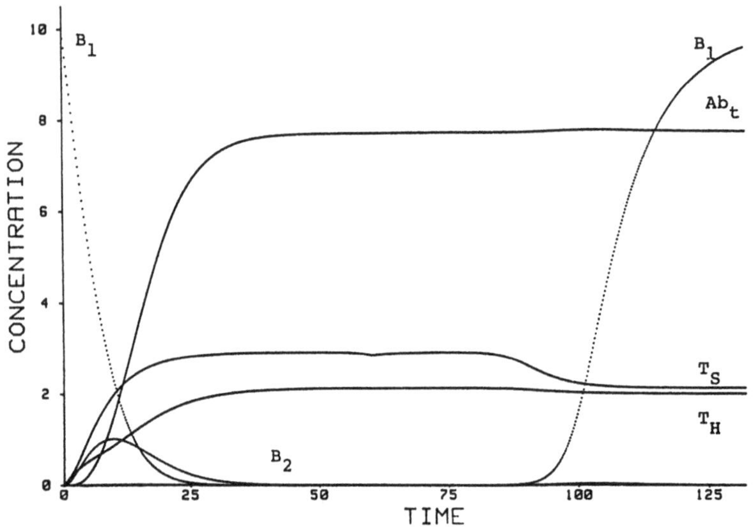

Figure 7 : High dose paralysis : no antibody can be produced for a period after pretreatment.

High dose paralysis is here not related to the behavior of the T cell core but results from the strong inhibition of immature B cells in the continuous presence of antigen. Recovery from paralysis occurs spontaneously and gradually. It reflects the generation time of new immunocompetent B lymphocytes in the absence of residual antigen. Our description, however, is not entirely satisfactory : too high antibody levels appear in the course of pretreatment, and T_S cells are not actively involved. These two aspects are related to the size of the primary response at high antigen concentrations as discussed hereabove.

4. Concluding remarks.

Our purpose has been to analyze a simple model of the immune response. The logical approach has been extremely useful for such a study, by itself and combined with a continuous approach. The generalisation of the discrete method moreover, improved strongly the predictability of the behavior of the continuous system from its logical caricature. A similar type of modelling in terms of binary automata has recently been developed by Atlan et al. for the dynamics of certain auto-immune diseases (see e.g. /14/).

The model is based on the concept that the immune system displays regulatory pathways in the absence of foreign antigen /15/. This leads to an interpretation of the various modes of the humoral response in terms of switches among multiple steady states. These transitions are determined by the strength of the antigenic stimulus, the way in which antigen is administered and the previous antigenic history of the system. The presence of positive loops is essential to observe a multiplicity of steady states but the details of the structure of these loops is not so important /10/.

Our description does not include a full characterization of all the cells and processes which are involved in the control of the immune response. In spite of severe simplifications it accounts however for a number of typical features.

Acknowledgements

This work is supported in part by the Belgian program on inter-university attration poles and in part by the C.E.E "Actions de Stimulations n° ST2-J-0385-C (GDF).

References

1. N.A. Mitchison : In Regulation and the Antibody Response, ed. by B. Cinader (Thomas Springfield, Illinois, 1969).
2. G.R. Shellam, G.J. Nossal : Immunology 14, 273 (1968).
3. M. Kaufman, J. Urbain, R. Thomas : J. Theor. Biol. 114, 527 (1985).

4. M. Kaufman, R. Thomas : J. Theor. Biol. 129, 141 (1987).
5. M. Kaufman : In Theoretical Immunology, ed. by A. Perelson Addison Wesley Pu. Co, Inc., 1988).
6. G. Nossal : Ann. Rev. Immunol. 1, 33 (1983).
7. D.R. Green, P.M. Flood, R.K. Gershon : Ann. Rev. Immunol. 1, 439 (1983).
8. G.K. Sim, I.A. McNeil, A.A. Augustin : Immunol. Rev. 90, 49 (1986).
9. G.W. Hoffmann : Contemp. Topics in Immunobiol. 11, 185 (1980).
10. M. Kaufman, J. Urbain : In Theoretical Models for Cell to Cell Signalling, ed. by A. Goldbeter (Academic Press, London, in press).
11. E.H. Snoussi : Dynamics and Stability of Systems, in press (1989).
12. R. Thomas, R. d'Ari : In Biological Feedbacks, (CRC Press, Boca Raton, Florida, in press).
13. P. Van Ham : In Dynamical Systems and Cellular Automata, ed. J. Demongeot, E. Golès, M. Tchuente (Academic Press, London, 1985).
14. G. Weisbuch, H. Atlan : J. Phys.A : Math. Gen. 21, L189 (1988).
15. N.K. Jerne : Ann. Immunol. (Inst. Pasteur) 125C, 373 (1974).

Simulation of the Immune Cellular Response by Small Neural Networks

H. Atlan and S. Hoffer-Snyder

Hadassah University Hospital, P.O. Box 12000, Jerusalem 91120, Israel

1.1 Neural Net Computation for Immunology

In previous works, (1,2) neural network computation has been proposed as a useful tool in simulating regulatory processes of autoimmunity. In the present communication, the principles of this method will be briefly summarized and an extension of the work presented, with the purpose of arriving at a generalization in regard to other aspects of the immune response.

As in actual neuronal networks, cooperative properties of memory, pattern recognition and learning are observed in immunology as a result of interactions between large numbers of cells. Therefore, as regards functional assemblies of neurons, these properties can be simulated by the dynamics of threshold automata networks. However, in place of synapses, the physical substrates for interactions are membrane- bound (or free) antigen-antibody recognition reactions. In addition, these interactions involve cell populations rather than individual cells. Thus, each automaton represents, in this case, a cell population assumed to be in an homogeneous state at each time interval unit. As usual in neural nets, computation is discrete in time and consists of the updating of the state of each automaton according to a transition rule, relating the state x_i of an automaton i at time t+1 to the inputs it has received from other connected automata j at time t.

More precisely:

$$x_i(t+1)=1 \quad \text{if} \quad \sum_j w_{ij}\, x_j(t) > \theta_i$$

$$x_i(t+1) = 0 \text{ otherwise} \tag{1}$$

where θ_i is a given threshold; w_{ij} is the weight of the connection from j to i and may be positive or negative, depending on the activating or suppressing nature of the connection. Changes in the states of cell populations are triggered by encounters with one another and with antigens. They consist in changes from a resting to an activated state which, in the case of B cells, involves the secretion of antibodies. Activation implies rapid cell division so that the population size changes from a few units to several powers of ten. In previous works, it was assumed as a first approximation, that the activation of cells and a significant increase in population size could be represented by one single (activated) state, rated as 1, the resting state being rated as 0. We assumed all the θ_i to be zero so that each time the state of a cell population is updated, it is given a state 1 if the algebraic sum of its inputs is larger than 0; otherwise it is given a state 0.

The different automata are updated (either synchronously, simulating a parallel computation, consecutively in an ordered series, or randomly) until an attractor has been reached in the form of a steady state or a limit cycle. It is clear that only small limit cycles are interesting, the cycle length of which is much shorter than the number of states which make up the whole state space. In cases where there is no attractor the network roams indefinitely in the whole state space. However, in such cases, since the network is finite, any deterministic order of updating will make it come back eventually to a previous state. Thus, deterministic updating in the absence of other attractors would always produce a trivial limit cycle made of the whole state space.

A given connection structure thus leads to one or several attractors which appear as the outcome of the assumed interactions between the different cell populations. It can be shown (3) that for any given structure, all the attractors, either steady states or limit cycles generated by random updating, are also generated by any other order of computation. The reciprocal is true only for steady states but not for limit cycles; in other words limit cycles can be generated by a given order of updating in parallel or in series and not by random updating. Therefore, the dynamics generated by random updating are the most general. Moreover, one can assume that in the absence of an external clock which would impose a synchronization, and of compartmentalization which would impose a specific temporal order in the interactions, encounters between cell populations occur randomly. That is why we feel safer in running the computation by randomly updating the automata.

For small networks a very simple algorithm allows one to scan the whole state space as possible initial conditions and to quickly find the attractors generated by such random updating.

In the absence of learning and with no specific information on the different strengths of the connections (namely the different affinities of receptor-ligand or antigen-antibody interactions) all the connection weights are given an absolute value of 1. Learning procedures consist of changing the connection weights according to previous experience. More specifically, a second order dynamics is generated by modification of the connections according to certain rules derived from the state of the network in its attractor. Thus, the network moves from this attractor and reaches a new one, which is produced by the new connection structure. We shall return later to the implementation of such learning procedures in simulating processes of vaccination.

1.2 Network Regulation of Autoimmunity

In our first applications of this kind of computation in collaboration with Irun Cohen and Gerard Weisbuch (1,2) we tried to simulate the regulation of autoimmunity as suggested by empirical data on Experimental Autoimmune Encephalomyelitis (EAE) (4).

In addition to effector T cells specifically recognizing the self-antigen (myelin basic protein) and directly responsible for the disease, two different sets of regulatory cells have been actually identified: i) antigen specific helper and suppressor T lymphocytes reactive to the antigen: it was postulated that the balance between their activating and suppressing effects on the effector cells is responsible for the regulation of autoimmunity (5). ii) antiidiotypic helper and suppressor T lymphocytes associated with the onset of resistance to EAE induced by T cell vaccination (4).

We asked ourselves the following questions: How are the interactions between the five cell populations (effectors and regulatory cells) responsible for the regulation of the autoimmune response, under normal conditions of tolerance, under pathological conditions of the onset of the disease, and under the conditions of acquired tolerance after a non-lethal form of the disease or after T cell vaccination? What are the functions of the two sets of regulatory T lymphocytes, since, a priori, one pair of helper and suppressor populations would seem to be sufficient to insure regulation?

We used a five automata network with different connection structures in order to simulate the effects of plausible kinds of interactions between the five T cell populations. Each automaton represented one of the cell populations as listed in table 1.

Table 1

Automaton:	I	II	III	IV	V
T lymphocyte population	Ag Specific Helpers and inducers	Effectors	Anti idiotypic suppressors	Anti-idiotypic helpers	Ag specific suppressors

The self antigen is assumed to exist in two possible forms: a suppressogenic form without adjuvant, assumed to activate only the antigen specific suppressors (cell population V); and an immunogenic form, where it is associated with an adjuvant which seems to be necessary for the activation of inducer T lymphocytes, (cell population I) in addition to the antigen specific suppressors . The effector cells (population II) are assumed to recognize the antigen under all circumstances, but this reaction is efficient only through the activation of cells I. That is why a direct connection from the antigen to cells II is not necessary in our network representations (fig. 1). In addition, the effect of the immunogenic antigen on cells II through the activation of cells I in the presence of adjuvant may be viewed as a schematic merging of two different functions which, in reality, pertain to two different, though related, types of cells: inducer T cells working as helpers and antigen-presenting cells.

In previous simulations (2) the antigen was represented by an external input of value 1 sent to automaton V only in the suppressogenic form, i.e. in the absence of adjuvant) and to V and I in its immunogenic form, (i.e. in the presence of adjuvant) (fig. 1).

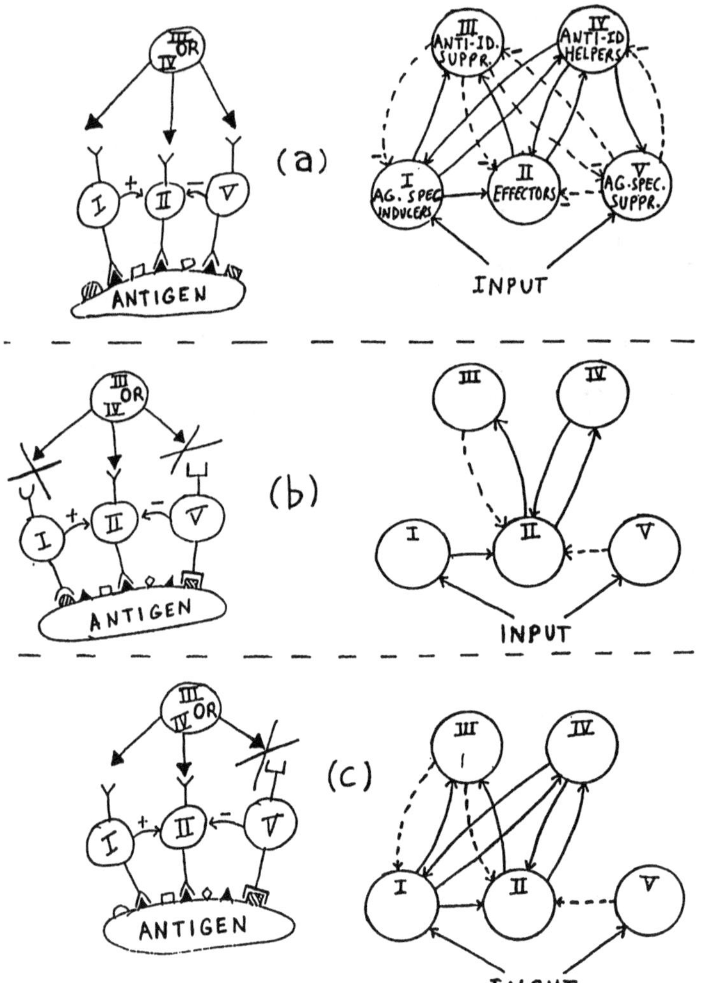

Fig. 1 Three connection structures are constructed on the basis of three different assumptions (refer. 2): (a) completely connected: antigen specific regulatory cells (I and V) recognize the same epitope as effector cells (II). Therefore they are recognized by the anti- idiotypic regulatory cells (III and IV). (b) separate connections: I and V recognize an epitope different from the one recognized by II. Therefore III and IV do not recognize I and V. (c) III and IV recognize I but not V.

<u>External inputs</u>: This model (2) is more detailed and more plausible than the first one presented in (1). An external input is sent from the self-antigen to cells I and V. In its suppressogenic form, the input to V only is set equal to 1; the input to I is set to 0. In the immunogenic form of the antigen, both inputs are equal to 1.
In section 2 where three-state automata are being used the antigen input is set equal to 2 on cells I and V to represent the immunogenic form, and to 0 on cells I and 2 on V to represent the suppressogenic form. Under certain assumptions, as indicated in

Different connection structures were tested, based on different assumptions on possible interactions between the two pairs of regulatory cells (fig. 1). As a first step, we did not assume learning procedures, but merely looked for the resulting attractors and their stability under modifications of the connections according to the different hypotheses. The attractors as all the $2^5=32$ possible states of the network are endowed with a biological meaning such as: naive, i.e. no previous acquaintance with the antigen (all the five automata are in 0 state); tolerant to an antigen, some of the helper and suppressor populations are activated but the effector is in 0); active against an antigen (the effector population is activated). One must notice that the cell populations can be activated "spontaneously" in the absence of antigen by random encounters with one another, if they receive positive connections, as well as by random fluctuations. That is why any one of all the 32 states must be considered as a possible initial state of the network, the attractor(s) being the only stable one(s). For example, the naive state as attractor does not imply that in the absence of antigen, no cell population can be activated, but only that when that happens the dynamics imposed by the connection structure bring all the cell populations back to a stable inactivated state. Fluctuations may take the network out of its attractor. It will always go back to it unless more than one attractor is generated by the dynamics. Then, it will go back either to the same or to another of the attractors, by chance, depending upon where in the state space (in which basin of attractors) it has been taken by the fluctuation.

In the next section, we shall mention conditions under which no attractor was observed, i.e. the network wanders randomly indefinitely in the state space. Even if "spontaneous" activation of every cell population can occur it has no chance of being stabilized and therefore this result will be interpreted as tolerance (see section 2).

The main results of our previous simulations can be summarized as follows:

Two different attractors were found in response to self antigen in its suppressogenic and immunogenic form respectively. In the first case, the state 00001, i.e. a state where cells I, II, III, IV, are in 0 and V in 1, was found as an attractor for the three

---

the text, an attenuated form of the antigen is represented by an input 1 to cells I; and a weakened reactivity of cells V is represented by an input 1 to automaton V. Normally, no direct activating connection is established between the antigen and the effector cells II (see text). In section 2.5 a model is presented where a stable diseased state is induced by a weakening of V associated with a direct attenuated activation of II. However, these pathological connections are sufficient to produce the disease, only when the antigen is in its immunogenic form (input 2 to I); when the antigen presenting system is attenuated (input 1 to I), the direct activation of cells II simulates the induction of a non-diseased state of activity by low doses of activated effector cells. As explained in the text, this state in its turn induces by learning a vaccination against a further exposure to the immunogenic antigen.

connection structures. With the antigen in its immunogenic form, an attractor 10001 was found in the complete and separate connection structures; with the partial connections the more interesting attractor 10111 was found where all the regulatory cells are activated but the effectors are kept inactivated. Thus, in both cases of antigen with and without adjuvant a stable state of self-tolerance is induced.

We proposed (2) that the existence of two different stable states of self-tolerance induced by the network in response to the two different forms of antigen would appear to represent two different levels of regulation discernible both experimentally and clinically. A first level would be mediated by antigen specific suppressors (V) that function to deactivate potentially virulent circulating effector cells (II). This level is proof against self-antigen without adjuvant, and it would be effective in healthy organisms where active effector T lymphocytes have been isolated (6,7,8) without being capable of producing disease spontaneously.

A second level of regulation protects against the antigen in its immunogenic form. It would be effective in the induced tolerance observed in rats after recovery from non-lethal forms of EAE and after T cell vaccination (4,9,10). This second, more advanced, level of resistance to activated effector cells is probably mediated by the anti-idiotypic T cells, particularly by anti-idiotypic suppressors (4,11).

Since some state of self tolerance was always found as an attractor, autoimmune disease would require some defect or change in the cells or connections of the network itself, such that any transient state of activity against the self antigen in its immunogenic form would not be driven back to a stable state of self tolerance. To ascertain how such pathogenic attractors might emerge, we weakened specific connections by assigning a value of "0" to each of the 4 regulatory cells in turn and tested the effect on the network response to immunogenic self-antigen. The absence of either automaton III or V enhanced the likelihood of autoimmune disease irrespective of which the 3 network connections were tested. However, the consequences were not identical, in that the absence of cells V (Ag specific suppressors) seems to be more effective in producing a stable state of activity (disease) than that of cells III (antiidiotypic suppressors). Since these results are qualitatively reproduced by the more extended model presented below, they will not be put forth in detail here.

The absence of the anti-idiotypic helpers (automaton IV) as could be expected, did not lead to loss of self-tolerance. However, as discussed elsewhere, the anti-idiotypic helpers perform a major function in regulating immunity to foreign antigens. Since distinction between foreign and self is acquired as a result of the developmental history of the immune system (as shown e.g. by neo-natal tolerance to foreign antigens "seen" as self), there is no reason to assume that the helper T lymphocytes, although not necessary for self-tolerance, would not develop in response to self as well as to foreign antigens. Indeed, as mentioned above, development of anti-idiotypic helper T cells was found to be induced, together with anti-idiotypic suppressors, by T cell vaccination against EAE.

2.1 The Three State Automata Network

Obviously the model had to be extended in order to incorporate at least two important phenomena: i) learning procedures must be used to represent vaccination as temporal evolution of the network structure. ii) the effector cells may have some destructive effect , not only on the antigen, but also on the antiidiotypic regulatory cells which reproduce, on their membrane, the structure of the antigen.

For this purpose, we now make explicit two different activated states rated 1 and 2 in addition to the resting state 0. They correspond to different stages in the modifications which occur during the processes of proliferation and differentiation which follow T cell activation. These changes in the cell state result in different effects of the cells on their environment.

The transition rules (1) of the automata now become:

$$x_i(t+1) = 0 \text{ if } \quad \sum_j w_{ij}\, x_j(t) \leq \theta_i$$

$$x_i(t+1) = 2 \text{ if } \quad \sum_j w_{ij}\, x_j(t) \geq \theta_i + 2$$

$$x_i(t+1) = 1 \quad \text{otherwise.} \tag{2}$$

As in the previous model, the thresholds θ_i are set to 0.

Obviously, the state space increases accordingly from $2^5=32$ to $3^5 = 243$ possible states for the network. Amongst them only those where cells II are in state 2 are considered immunologically active against the antigen, i.e. diseased in the case of self antigens. The three state automata network allows us to incorporate the regulatory effects of effector cells into the model. As mentioned above, the effector cells have an activating effect on the antiidiotypic cells which recognize their idiotype but they also have a suppressive effect on the same cells which predominates when they have reached their second stage of activity. This effect is represented by changing the sign of the signals sent by automaton II to III and IV according to the state for II: from +1 when II is in state 1 the signals become -1 when II is in state 2. (This phenomenon would be expressed in continuous kinetics by an antiidiotypic cell activity increasing first and then decreasing as a function of effector concentration).

2.2 Differences in the Protective Effects of the Two Kinds of Suppressor Cells

The results of the simulations were qualitatively similar to those reported in section 1: the attractors produced by both the complete and separate connection structures, were 00002 and 20002., with the antigen in its suppressogenic and immunogenic form respectively. As previously, the partially connected structure produced the more interesting attractor 20222 with the antigen in the presence of adjuvant (table 2).

Table 2

Connection Structure	Form of Self-Antigen	Antigen input value	Attractors
Complete	Suppressogenic	0	00002
Separate	Immunogenic	2	20002
Partial	Suppressogenic	0	00002
	Immunogenic	2	20222

In other words, tolerance is always produced by the 5 cell network. The "deletions" of regulatory cells also produced effects qualitatively similar to those previously observed with the two state automaton networks (table 3). Deletion of cells IV resulted in the attractor 20002 with the complete and separate connection and in a cycle made of the 4 tolerant states 00002--->10002 ---

Table 3
Loss of tolerance to the immunogenic self-antigen by "deletion" of regulatory cells

Connection Structure	Cell population deleted	Attractors
Complete	None	20002
	III	20002 and 21012
	IV	20002
	V	22110
Separate	None	20002
	III	20002 and 21012
	IV	20002
	V	22000
Partial	None	20222
	III	cycle: 22022-->22012-->21012-->21022
	IV	cycle: 00002-->10002-->10102-->00102
	V	22110

>10102--->00102 with the partial connections. Deletion of the suppressor cells produced interesting results in that the effects of cells III and V appear to differ significantly. Deletion of cells V always resulted in a complete and stable loss of tolerance, i.e. a stable attractor where cells II are in state 2, namely 22110 with the complete and partial connections and 22000 with the separate connections. On the other hand deletion of III with the complete and separate connections produced the <u>two</u> inactive attractors 20002 and 21012, (in the latter, cells II are activated but not enough to produce the disease). It is only with the partial connections that deletion of cells III produced a loss of tolerance, but still unstable: the four state cycling attractor 22022--->22012--->21012--->21022, so that the network oscillates between two states of disease (cells II in state 2) and two states of active tolerance (i.e. II in state 1).

In other words, providing that the antigen specific suppressors (V) are effective, antiidiotypic suppressors (III) do not seem to be absolutely necessary to protect the system against the disease, even with the antigen in its immunogenic form. However, it is clear that they have a definite protecting effect under abnormal conditions where, due to an imbalance in the first set of regulatory cells (antigen specific), the disease may be induced. This protective effect appears even more clearly if we try to simulate a learning process of acquired tolerance induced after the onset of a non lethal disease or after T cell vaccination.

2.3 <u>Learning as Second Order Dynamics</u>

Delocalized memory and learning are general performances of neural networks (12,13). A possible mechanism for their implementation is the following: an attractor (or a set of attractors) represents the response to a given connection structure between the elements of a network. This response itself, as a first experience, may induce changes into the connections which in their turn will produce new behaviors with new attractors. Thus previous experience is stored in the form of connection weights and produces a new, learned, behavior in the form of new attractor(s).

It is worth noting how big the connection space is, even for small networks. The number of different ways to connect N automata is 2^{N^2} if only two possible weights 1 and 0 are assumed for each connection, i.e. connected and non-connected. If p is the number of possible weights, this number is p^{N^2}. There exists a high degree of "underdetermination of the models by the facts" (14) in the sense that large classes of different connection structures produce the same attractors. Several learning rules have been proposed to change the connection weights, depending upon the context of applications (12,13). Therefore, the rules we choose must rely on plausible biological assumptions and the model may be used as a possible falsifying test for these assumptions rather than a verification.

2.4 <u>Acquired Tolerance After Disease or Vaccination</u>

We have seen that the loss of tolerance is steadily simulated by the model when we assume that antigen specific suppressors are absent or ineffective (automaton V deleted). Such an incomplete network can be used to simulate the acquisition of tolerance from

Table 4

New attractors after application of the learning rule* to the network with automaton V deleted.

Connection structure		Form of self-antigen**	Attractors
Complete	Before learning	Suppressogenic	00000
		Immunogenic	22110
		Attenuated	11220
	After learning	Suppressogenic	00000 and 11220
		Immunogenic	no attractor
		Attenuated	no attractor
Separate	Before learning	Suppressogenic	00000
		Immunogenic	22000
		Attenuated	11110
	After learning	Suppressogenic	00000 and 11110
		Immunogenic	no attractor
		Attenuated	no attractor
Partial	Before learning	Suppressogenic	00000
		Immunogenic	22110
		Attenuated	11220
	After learning	Suppressogenic	00000 and 11220
		Immunogenic	no attractor
		Attenuated	no attractor

* The learning rule consists of changing the connection weight w_{I-II} from 0 to 1 when an attractor has been reached where automaton II is in state 1 or 2. Note that all connections from II change their sign from +1 to -1 during the computation (the first order dynamics) when II is in state 2 (see text).

** Suppressogenic, immunogenic and attenuated mean input value on I respectively 0,2,1 (see legend to fig. 1). (Input value on V is always 0 since V is deleted).

a state of disease by learning from experience (see table 4). We start from the state 22110 produced by deletion of V in the complete and partial connection structures. (Similar results are obtained from 22000 in the case of separate connections).

As a learning rule we utilize a very simple one which relies on the following assumption: before learning there is no connection from II to I, meaning that as long as cells II have not been activated by a particular antigen they have neither an activating nor a suppressing effect on the automaton I which represents, as mentioned above, both antigen specific helpers and the antigen presenting system. However, as soon as they are activated in a stable manner, (i.e. they are stabilized in state 1 or 2 in an attractor of the network), one can assume that they interact with cells I as they do with antiidiotypic cells III and IV. The very fact of their activation makes them react with the antigen and consequently with the antigen presenting system. Thus, a new connection is established from II to I. Some evidence has been presented (15) to support the plausibility of this phenomenon. When we apply this rule we find it sufficient to account for the process of acquired tolerance both after autoimmune disease and after T cell vaccination.

As mentioned above, we assume that the stable state of disease is represented by the attractor 22110 (or 22000) produced by deletion of automaton V. By application of the learning rule a connection is established from II to I, since cells II are activated in the attractor. This results in a new network structure. In the absence of adjuvant the new network now generates two attractors 00000 and 11220 with complete and partial connections, (00000 and 11110 with the separate connection structure).

The interesting result is that now in the presence of adjuvant the network does not stabilize in any attractor. In other words the onset of the disease by the immunogenic antigen has changed the network structure in such a way that additional exposure to the antigen in the same immunogenic form does not produce a stable state of disease.

A similar result is obtained by exposure of the network to an attenuated form of the immunogenic antigen (table 4). This can be simulated by giving a value 1 instead of 2 to the antigen input to cell I. In the absence of learning by previous exposure (still with V deleted) the attractor is 11220 for the complete and partial connections, and 11110 for separate connections. In other words a stable non-diseased state is established where effector cells II are still activated but prevented from actively dividing by the effect of antiidiotypic suppressors cells III. Therefore, by application of the learning rule a new connection is established from II to I.

Then, for the three connection structures, the new dynamics do not exhibit any attractor, not only in response to the attenuated antigen but also to the immunogenic antigen in its full strength. In other words, the network has been vaccinated against the disease by the previous exposure to the attenuated immunogenic antigen.

2.5 Direct Activation of the Effector Cells

The 3-state automata network allows one to simulate an autoimmune diseased state partly determined by a direct activation of the effector cells (II) (an input of value 1 is directly sent to automaton II from the antigen (see legend to fig. 1)). This is meant to represent a genetic susceptibility to the disease in the form of a lymphocyte membrane receptor directly reactive to the myelin BP. However, this modification of the network by itself is not sufficient to produce a diseased state as an attractor. Some weakening of the suppressive effect of cells V is still necessary, although, contrary to the previous models, this effect does not have to be totally absent. Such a weakening is represented by an antigen input value of 1 instead of 2 to automaton V. Then, with the antigen in its immunogenic form the disease state 22001 or 22111 is produced as a single attractor (table 5). Applying the learning rule of establishing a connection from II to I produces an acquired tolerance again (no attractor).

Table 5

Vaccination against the autoimmune disease produced by direct activation of the effector cells associated with a weakening of the suppressive effect of automaton V.*

Connection Structure		Form of self-antigen**	Attractors
Complete	Before learning	Suppressogenic	00001
		Immunogenic	22001
		Attenuated	11111
	After learning	Suppressogenic	00001 & 11111
		Immunogenic	no attractor
		Attenuated	no attractor
Separate	Before learning	Suppressogenic	00001
		Immunogenic	22001
		Attenuated	11111
	After learning	Suppressogenic	00001 & 11111
		Immunogenic	no attractor
		Attenuated	no attractor
Partial	Before learning	Suppressogenic	00001
		Immunogenic	22111
		Attenuated	11221
	After learning	Suppressogenic	00001 & 11221
		Immunogenic	no attractor
		Attenuated	no attractor

* The learning rule is the same as in table 4; ** defined as in table 4.

The T cell vaccination observed after injection of non pathogenic doses of activated effectors is also easily simulated by the 1 input into cells II, together with input 1 to cells V.

As in paragraph 2.4 the network is exposed to an attenuated form of the immunogenic antigen, i.e. an input 1 instead of 2 is sent to cells I. The resulting attractor is 11111 or 11221 (table 5). Thus, cells II are in a stable activated state although not sufficiently to produce the disease. Therefore, the learning rule applies and the learned network reacts to the antigen even in its strong immunogenic form (input 2) by acquired tolerance (no attractor).

It is worth noting that a strong effect of antigen specific suppressors in this case (input 2 to cells V) would be counterproductive as far as this vaccination is concerned. The attractor produced by the attenuated antigen in this case is such that the learning rule cannot be applied because cells II are in state 0 (10112 for the partial connections and 10002 for the two other connection structures). In other words the suppressor effect is too strong to allow for the minimum activation of the effectors required to induce learning. On the other hand, no learning is necessary in this case to prevent the disease, since the full suppressive activity of V (and III) maintains cells II in state 1 even when the immunogenic antigen is present.

3. Conclusion: Foreign Versus Self Antigens

The same Five-automaton unit may serve as a model for cellular immune response to a foreign antigen. The basic difference with autoimmunity consists of a weak or lacking effect of one of the suppressor cell populations, as the normal state of affairs. Moreover, not only the antigen specific suppressors (V) but also the antiidiotypic suppressors (III) may be used for this purpose with, obviously, a different biological meaning.

In addition, the learning rule induced by a first encounter with the foreign antigen is different, since it leads to new attractors with more activity of the effectors rather than more tolerance.

A plausible rule is derived from the phenomenon of affinity maturation and is an equivalent for the immune system to Hebb's rule of synapse reinforcement for the central nervous system. However, since somatic mutations in T cells do not seem to happen as frequently as in B cells, we should consider a more complete network, involving B cells and antigen-antibody reactions, to account for responses to foreign antigens where affinity maturation would play an important role. This point is the topic of an ongoing work and we cannot elaborate on it in the present paper.

However, the results presented here are sufficient to show similarities and differences between cellular responses to self and to foreign antigens. One main difference consists of a weaker effect of antigen specific suppressors in the normal response to foreign antigens. This difference does not have to be related to intrinsic molecular properties differentiating self from non-self structures. It may be the result of past history of encounters with subsequent antigenic structures during the maturation of the whole immune system. We wish to suggest a suppressor saturation phenomenon whereby the normal development of regulatory networks of the kind suggested here could take place only during a limited period of time in maturation. Then, (thymic?) capabilities to produce new suppressor cell populations would be exhausted. Thus, the antigen structures which were present before the saturation appear to be defined as self a posteriori as it is suggested by neonatal tolerance of genetically foreign antigens.

ACKNOWLEDGEMENTS

We wish to thank I.R. Cohen, G. Weisbuch, A. Coutinho and F. Jacquemard for their wise and useful advice on several occasions.

This work was partly supported by INSERM (Paris) grant 87/9002.

REFERENCES

1. G. Weisbuch and H. Atlan: In J. Phys A.: Math. Gen. 21, 189 (1988)

2. I.R. Cohen and H. Atlan: In J. Autoimmunity, 1989, in press.

3. H. Atlan, I. Cohen and G. Weisbuch, in preparation.

4. O. Lider, T. Reshef, E. Beraud, A. Ben-Nun and I.R. Cohen: In Science 239, 181 (1988)

5. E.E. Sercarz, R.L. Yowell, D. Turkin, A. Miller, B.A. Araneo and L. Adorni: In Immun. Rev. 39, 108 (1978)

6. H.J. Schluesener and H. Wekerle: In J. Immunol. 135, 3128 (1985)

7. S. Orgad and I.R. Cohen: In Science 183, 1083 (1974)

8. J.A. Burnes, B. Rosenzweig and R.P. Lisak: In Cell Immunol. 81, 435 (1983)

9. I.R. Cohen: In Immunoregulatory Processes in Experimental Allergic Encephalomyelitis and Multiple Sclerosis ed. by A.A. Vandenbark and J.C.M. Raus 7, 91 (Elsevier, Amsterdam 1984)

10. A. Ben-Nun and I.R. Cohen: In J. Immunol. 128, 1450 (1982)

11. D. Sun, Y. Quin, J. Chluba, J.T. Epplen and H. Wekerle: In Nature 332, 843 (1988)

12. G. Weisbuch: In J. Theoret. Biol. 121, 255 (1986)

13. D.E. Rumelhart and J.L. McClelland: In Parallel Distributed Processing, Vol. 1 (MIT Press, Cambridge Mass. 1986)

14. H. Atlan: In Bull. Mathem. Biol. (1989) in press.

15. A. Ben-Nun and I.R. Cohen: In Eur. J. Immunol., 12, 709 (1982)

Discrete Time Versus Continuous Time Approach to the Autoimmune Response

K.E. Kürten

Institut für Zoologie III, Biophysik, Johannes Gutenberg-Universität, D-6500 Mainz, Fed. Rep. of Germany and
Institut für Theoretische Physik, Universität zu Köln, Zülpicher Straße 77, D-5000 Köln, Fed. Rep. of Germany

A discrete time model for the immune regulation recently proposed by Weisbuch and Atlan is extended to continuous coupling coefficients as well as to continuous concentration rates in continuous time. It is shown that depending on the choice of the network parameters typical immune responses can be observed.

1. INTRODUCTION

When a pathogen invades and challenges an organism, the immune system responds in two fundamentally different ways, by humoral immunity and by cellular immunity. The second type of immune reaction modelled in this report is based on a cell population collectively known as T lymphocytes which originate in the bone marrow but complete their development by maturation in the thymus gland. T cells can either help or suppress the humoral response to a stimulus, thus regulating the activity of the B cells which compose the humoral system. Three fundamentally different types of T lymphocytes are recognized: killer, helper and suppressor cells each of them having many subdivisions. According to the postulates of the "clonal selection theory" [1], a recognition process provides the organism with active killer T cells able to destroy cells that display the foreign protein. While the appropriate epitope is detected, the lymphocyte is stimulated to divide reproducing more lymphocytes or to secrete free antibodies.

2. FORMULATION IN DISCRETE TIME

2.1 The model

Recently Weisbuch and Atlan proposed a fairly idealized model for the auto-immune response with a view to producing a system that is simple enough to be described by a binary automaton network but still complex enough to embody some important characteristics of the immune regulation [2]. Guided by Cohen's experimental findings on autoimmune encephalomyelitis [3] and in close connection with Jerne's postulate [1] they presented a network of five interconnected pools consisting of binary formal threshold cells: killer cells C_1, activated killer cells C_2, helper cells C_4 and two kinds of suppressor cells C_3 and C_5. The state of each pool can either take the value $\sigma_i = 1$ or $\sigma_i = 0$, $i = 1, \ldots, 5$, depending on whether a cell type is present in a high or low concentration. The synchronous and fully deterministic dynamics of this network reminiscent of the pioneering discrete time neural network model of McCulloch and Pitts [4] is described by the following set of equations, where the five states are supposed to be updated in parallel according to

$$\sigma_i(t+\tau) = \Theta\left(\sum_{j=1}^{5} c_{ij}\, \sigma_j(t)\right) \quad i = 1, \ldots, 5 \tag{1}$$

The theta function is set to unity for positive arguments and zero elsewhere. The quantity τ can be related to the scale of the incubation time presumably lasting a few days. The topology of the network is specified by the coupling matrix C defined as follows:

$$C = \begin{vmatrix} c_{11} & 0 & -c_{13} & c_{14} & 0 \\ c_{21} & 0 & -c_{23} & c_{24} & -c_{25} \\ c_{31} & 0 & 0 & 0 & 0 \\ c_{41} & 0 & 0 & 0 & 0 \\ 0 & 0 & 0 & c_{54} & 0 \end{vmatrix} \tag{2}$$

Without loss of generality the strengths of the interactions between the cells specified by the quantities c_{ij} can be chosen arbitrarily in the interval [0,1]. Positive (negative) signs of the c_{ij}s reflect excitatory (inhibitory) effects of pool j on pool i, zeros mean that there is no interaction between pool j and pool i. Note that the coupling coefficients playing the role of the synaptic strengths in neural network models are in general not symmetric in contrast to the immune network recently proposed by Hoffmann.[5]

2.2 Dynamical properties

It has already been shown in ref.[6] that the problem can essentially be reduced to the two-pool system with the dynamical evolution

$$\sigma_1(t+\tau) = \Theta\{\sigma_1(t) + x_1\sigma_3(t)\}$$
$$\sigma_3(t+\tau) = \Theta\{\sigma_1(t)\} \qquad (3)$$

The sign of the control parameter $x_1 = (c_{14}-c_{13})/c_{11}$ describes the balance of the inhibitory and excitatory efficacies with respect to the killer pool C_1, large magnitudes of x_1 reflect a low degree of autocatalytic proliferation. One can show further that irrespective of the choice of the coupling coefficients the states of all pools will eventually be identical except that of the active killer cells. The relevant state of this pool can then finally be determined from

$$\sigma_2(t+\tau) = \Theta\{x_2\sigma_1(t)\}\ . \qquad (4)$$

The coefficient

$$x_2 = c_{21} + c_{24} - (c_{23} + c_{25}) \qquad (5)$$

serves as a second control parameter describing the balance of the strength of inhibitory and excitatory effects with respect to the active killer pool C_2. Thus, only if $x_2 > 0$, when the excitatory efficacies are prevalent, the pool of the active killer cells C_2 can be in state one. To ease classification the "number" of an attractor with respect to its configuration is defined in terms of its decimal representation according to [2]

$$\underline{n} = \sigma_1 + 2\sigma_2 + 2^2\sigma_3 + 2^3\sigma_4 + 2^4\sigma_5 \qquad (6)$$

varying between $\underline{0}$ and $\underline{31}$.

Equation (3) implies that the model exhibits three stable fixed points: the healthy naive state $\underline{0}$=(0,0,0,0,0), the healthy carrier state $\underline{29}$=(1,0,1,1,1) and the totally infected state $\underline{31}$=(1,1,1,1,1). Table 1 shows which attractors are reached from different initial concentrations of the pools C_1, C_3 and C_4 as a function of the control parameters x_1 and x_2. The initial concentrations of the pools C_2 and C_5 can be chosen arbitrarily.

The model proposed by Weisbuch and Atlan [2], where the quantities x_1 and x_2 are strictly zero, exhibits only the two attractors $\underline{29}$ and $\underline{0}$ with basins of attraction of equal magnitude depending on the dominance of excitatory or inhibitory effects in the initial concentrations. Assuming that autoimmune disease is caused by states in which the active killer cells are present in high concentrations there is always a hundred percent chance of surviving, if the incubation time for intermediate states like $\underline{31}$ is not too long. For the model generalized to

Table 1: Stable attractors corresponding to different initial concentrations for different control parameter regimes.

C_1	C_3	C_4	attractor	control parameter regimes
1	1	1		
			31	$x_1 > -1$ and $x_2 > 0$
1	0	1		
			29	$x_1 > -1$ and $x_2 \leq 0$
0	0	1		
			0	$x_1 \leq -1$
1	0	0		
			31	$x_1 > 0$ and $x_2 > 0$
0	1	1		
			29	$x_1 > 0$ and $x_2 \leq 0$
1	1	0		
			0	$x_1 \leq 0$
0	0	0		
			0	
0	1	0		

continuous coupling coefficients sampled randomly from a uniform distribution one can easily show that the attractor 0 is reached by about half of the initial conditions , whereas the attractors 29 and 31 equally share the remainder. Take into account that the appearance of attractor 31 inevitably causes death of the organism.

3. FORMULATION IN CONTINUOUS TIME

3.1 The model

Lymphocytes operate asynchronously and biological variables are usually continuous so that differential rather than finite difference equations provide a more refined and quantitative biological analysis. In this sense, an alternative description of the problem, based on the same network topology, can be formulated in continuous time by introducing nonlinear differential equations, thus breaking the artifical synchronism. The natural modification of ansatz (1) via differential equations suggests a sigmoid *continuous* nonlinearity instead of the Heaviside step function. Then, the set of five coupled non-linear first-order differential equations governing the dynamics are

$$a\dot{x}_i(t) = -x_i(t) + \xi\Big(\sum_j^5 c_{ij}\, x_j(t)\Big) \qquad i = 1,\ldots,5 \tag{7}$$

Here, the amount of change of the normalized *continuous* concentration rate of pool i, $x_i(t)$, is given by the difference between the proliferation rate and the spontaneous decay rate. The sigmoid function $\xi(u)$,chosen as $\xi(u)=(1 + \exp(-u))^{-1}$, takes care of the decreased sensitivity of the immune response at large excitatory or inhibitory inputs, the parameter a describes a time constant.

In analogy to the discrete time model the equations of motion for pool C_2 and pool C_5 decouple. Thus, the dynamical evolution of the system depends only on the five coupling coefficients $c_{11}, c_{13}, c_{14}, c_{31}$ and c_{41}. The steady state solutions of the relevant subsystem are then given by

$$x_1 = \xi(c_{11}x_1 - c_{13}x_3 + c_{14}x_4) \tag{8.1}$$

$$x_3 = \xi(c_{31}x_1) \tag{8.2}$$

$$x_4 = \xi(c_{41}x_1) \tag{8.3}$$

Substituting (8.2) and (8.3) in (8.1) this set of equations reduces to a one component transcendental equation in x_1

$$x_1 = \xi\{c_{11}x_1 - c_{13}\xi(c_{31}x_1) + c_{14}\xi(c_{41}x_1)\}\,. \tag{9}$$

Standard stability analysis leads to the characteristic equation with the solutions

$$\lambda_1 = -1$$

$$\lambda_{2/3} = -1 + \frac{c_{11}x_1(1-x_1)}{2}\Big\{1 \pm \sqrt{1 + \Big(\frac{2}{c_{11}x_1(1-x_1)}\Big)^2 \{c_{41}c_{14}x_4(1-x_4) - c_{31}c_{13}x_3(1-x_3)\}}\Big\} \tag{10}$$

For coupling coefficients of equal magnitude $|c_{ij}| = c$ the fixed point equation (10) simplifies to

$$x_1 = \xi(cx_1) \tag{11}$$

Thus, the continuous time model with couplings of equal magnitude exhibits only one single fixed point $(x_1^*, 0.5, x_1^*, x_1^*, x_1^*)$ corresponding to the fixed point $\underline{29} = (1,0,1,1,1)$ in the discrete time model. This fixed point is always stable with $0.5 \leq x_1^* < 1$ and x_1^* appoaches its limiting value one with increasing c. This stands in marked contrast to the discrete time model, where two stable fixed points with basins of attraction of equal magnitude have been found. For the model with arbitrary coupling coefficients computer simulations reveal up to three fixed points from which up to two are stable.

3.2 Application

Let us assume that, in spite of the technical difficulties, a fixed point $\underline{x}^*$ of eq.(7) can be determined experimentally through a sufficiently large period of observations. In general, the special topological structure of the model then allows the determination of the coupling coefficients c_{ij}. Equation (8) can be written as

$$c_{11}x_1 - c_{13}x_3 + c_{14}x_4 = \ln \frac{x_1}{1-x_1} \tag{12.1}$$

$$c_{31}x_1 = \ln \frac{x_3}{1-x_3} \tag{12.2}$$

$$c_{41}x_1 = \ln \frac{x_4}{1-x_4} \tag{12.3}$$

For a given fixed point $\underline{x}^*$ the coupling coefficients c_{31} and c_{41} are uniquely determined by eqs. (12.2) and (12.3). The quantities c_{13}, c_{14} and c_{11}, satisfying (12.1), can serve as suitable parameters so as to guarantee its stability. Thus, c_{11} can be adjusted to determine the sign of the real part of the eigenvalues, whereas suitable choices of c_{41} and c_{31} might produce complex eigenvalues, necessary for the existence of damped or sustained oscillations. The other sub-system, the active killer pool C_2 and the helper pool C_5 can easily be treated after the dynamics of the system has been determined. In analogy to (12) the fixed point equation takes the form

$$c_{21}x_1 - c_{23}x_3 + c_{24}x_4 - c_{25}x_5 = \ln \frac{x_2}{1-x_2} \tag{13.1}$$

$$c_{54}x_4 = \ln \frac{x_5}{1-x_5} \tag{13.2}$$

The coupling coefficient c_{54} can be determined from (13.2), whereas c_{21}, c_{23}, c_{24} and c_{25} serve as free

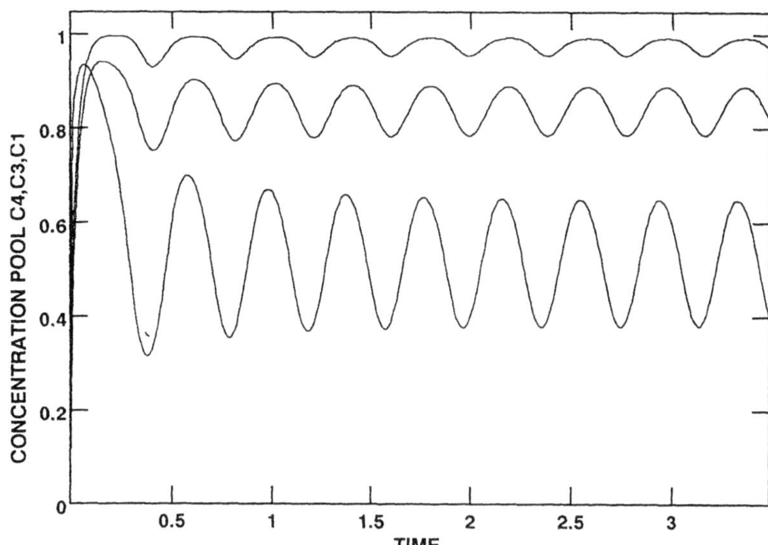

Fig.1: Illustrative solution of equations of motion. Concentration rates vs time for the killer pool C_4, the suppressor and helper pools C_3 and C_1 (from top to bottom) with coupling coefficients $c_{11}=8.2$, $c_{13}=15.43$, $c_{14}=9.$, $c_{31}=3.27$ and $c_{41}=7.74$. The time scale according to the time constant a can be chosen arbitrarily.

parameters to satisfy (13.1). Figure 1 shows a representative example for the appearance of cyclic behavior frequently observed in the overall immune response. When such sustained oscillations are present, all pools oscillate with exactly the same frequency. It is conceivable that these oscillations point to the concept of periodic or dynamic diseases which might arise from bifurcations in the dynamics of the physiological control system [7]. Note that oscillatory behavior, where the system undergoes a see-saw motion between tolerance and disease, does *not* occur in the discrete time formulation.

4. CONCLUSION

The results show that the evolution and function of our model network depends crucially on the excitatory and inhibitory balance between the connectivity parameters representing the key for self-tolerance or disease. However, in analogy to neural network models the coupling coefficients are time dependent underlying learning and self-organization processes. Following this direction the basic models might be extended to include external signals of antigens and to incorporate activity-induced plastic coupling coefficients enabling the network to respond more properly to its environment. Furthermore, one could model the time evolution of an antigen pool - more naturally incorporated in a continuous time description - and study primary and secondary responses for various ranges of antigen doses as well as the disappearance of high concentrations in the antigen pool [8].

In summary, our findings suggest strongly that this over-simplified model is a pragmatic but still limited approach to a complex problem, where the continuous time version more broadly complements the basic model in discrete time. However, this attempt might nevertheless serve as a reasonable starting point constituting a new direction for a network approach to the immune system via binary cellular automata operating in discrete time.

Acknowledgement

The author likes to thank U. an der Heiden, G. Weisbuch and M. Husson for helpful discussions. Funding for this work was provided by the Deutsche Forschungsgemeinschaft under grant Se 251/30-1.

References:

1. N.K. Jerne, Sc.Am. 229,1,52 (1973)
2. G. Weisbuch and H.Atlan, J.Phys. A21,189 (1988)
3. I.R. Cohen, Immunological Reviews,94,5 (1986)
4. W.S. McCulloch and W.H. Pitts, Bull. Math. Biophys., 5,115 (1943)
5. G.W. Hoffmann, J. theor. Biol., in press.
6. K.E. Kürten, J.Stat.Phys. 52,1/2 489 (1988)
7. M.C. Mackey and U. an der Heiden, Funkt.Biol.Med I,156 (1982)
8. M. Kauffman and R. Thomas, J.theor.Biol. 129,141 (1987)

Optimizing the Immune Control of Parasitic Invasion

Z. Agur

Dept. of Applied Math. & Computer Science,
The Weizmann Institute of Science, Rehovot 76100, Israel

The aim of this work is to identify the crucial tasks of the immune system in controlling parasitic invasion and to examine plausible molecular mechanisms by which these tasks may be effectuated. To this end the *Sleeping Sickness*, and its causative agent *Trypanosoma brucei* are discussed in terms of a general optimization problem, thus illustrating the major difficulties in controlling this disease. Subsequently an automata model is presented, of B-cell population dynamics under the constraints of a trypanosome-like pathogen. The efficacy of different mechanisms for optimizing the immune response are tested.

1. Introduction

Many scientific problems can be formulated as an optimization problem in which some cost function must be minimized. Such problems can be represented by a two-dimensional landscape of the cost function $f(x) = f(x_1, x_2, \ldots, x_N)$, which has many local minima (Fig. 1). In a simple description of the immune control as an optimization problem $f(x)$ will be the damage to the host and x will be an N-dimensional vector defining the linear genomic information encoding the antibody.

It has been demonstrated that antigen–antibody binding depends on recognition of topographical structures which, in turn, depend on conformation of the native molecule. As a result paratopes that accommodate large genomic sequence variations may preserve the essential secondary structural elements and their interactions, and vice versa, small differences in nucleotides may largely alter the paratope–epitope binding affinity [1]. This major property of antigen–antibody binding suggests that the immune landscape must have many local minima. Solving the optimization problem involves a rapid location of a reasonably low-cost minimum, corresponding to the antibodies that would most efficiently bind to the antigen and, hence, will maximize their proliferation rate.

This optimization problem can be compared with combinatorial problems such as the travelling salesman, that are NP complete, but to which a reasonably good solution can be

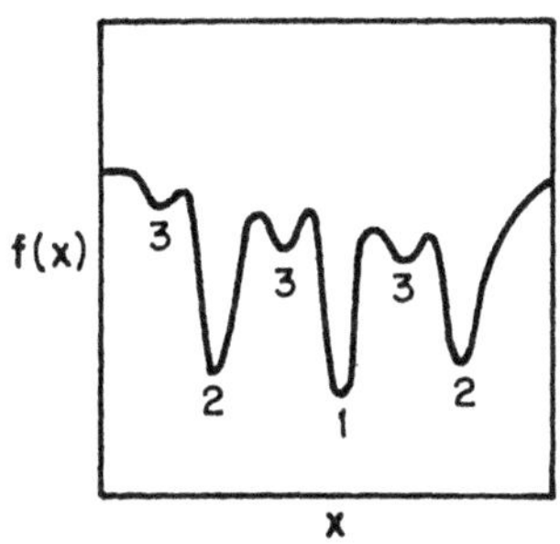

Fig. 1. A two-dimensional representation of a cost function $f(X) = f(x_1, x_2, \ldots, x_N)$ which has many local minima (2) with values of f close to the global minimum values (1). Adapted from [2].

obtained [2]. For a more precise illustration of the immune control optimization problem we will relate to a specific host-pathogen interaction, namely that of the Sleeping Sickness, trypanosomiasis.

Human Sleeping Sickness is caused by a subspecies of the African trypanosome *Trypanosoma brucei.* This parasite escapes immune-destruction through the periodic generation of a few daughter cells in which the antigenic composition of the cell surface coat is entirely altered [3,4,5]. The protective cell surface coat of each trypanosome consists of about ten million closely packed, identical variant cell surface glycoprotein (VSG) molecules. Each coat is encoded by a single VSG gene and by expressing a new VSG gene, out of a repertoire of about 1000 such genes, trypanosomes can periodically change coats and escape immune-attack [6,7,8]. Parasitaemias have a characteristic pattern with gradually increasing parasite numbers, whose majority express the same VSG on their coat — the *major VSG* [9]. The increase in parasite number is followed by increasing antibody titres against the major VSG. This humoral antibody response leads to the decimation of the parasite population, which is then replaced by the next antigenic variant (Fig. 2) [10,11].

Preliminary evidence indicates that VSG gene switching is an overall random process [13,14]. Nevertheless the sequence of dominant VSGs in relapse populations seems to have a "statistically definable order" [15]. This order has been explained in a mathematical model by the existence of double expressor switch intermediates with variable susceptibility to the immune response [16].

Interestingly, there is evidence that the damage in trypanosomiasis is primarily due to hyper-inflammatory reactions and disorder of the host's immune response [9]. That the immune system is "confused" in its attempts to solve the optimization problem indicates the special difficulty involved. If the control of a single parasitaemia wave can be described by a two-dimensional landscape, then successive parasitaemia waves should be represented by many different such two-dimensional landscapes. The additional time dimension to the landscape further complicates the optimization problem, especially since the transitions on the time axis are so crucial.

It is evident from the above description that in order to solve this three-dimensional problem the immune control should consist of well-timed fluctuations between a state in which a large antibody repertoire is being "screened" (i.e., formed), and a state of high fidelity in the production of a small subset of antibodies with best binding affinity to the major VSG. Less clear are the methods by which these frequent fluctuations between variability and resilience

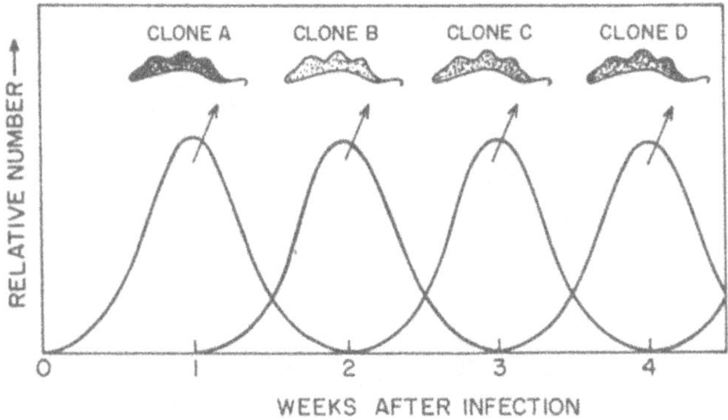

Fig. 2. Fluctuating parasitaemia of the Sleeping Sickness – trypanosomiasis. Each parasitaemia wave corresponds to a distinct major VSG (adapted from [12].

can most effectively be carried out. This problem is investigated below in an automata model, which tries to capture the most important properties of biological information processing and antigen–antibody interactions. The assumptions of this model are completely independent of the optimization landscapes discussed above, although its results can be used for drawing such landscapes.

2. The Model

In order to study the efficacy of various mechanisms for optimizing the immune control we employ an automata model of a B-lymphocyte population. In our model the genome of each *B-lymphocyte* is described by a binary string of n bits. This information is processed to produce a new n-bit string, which is further processed to produce a third string, and so on to level m, which is the final phenotype — the antibody. Let $a_{i-1,j-r}, a_{i-1,j-r+1}, \dots, a_{i-1,j+r}$ be the input values to the element located in the i-th row and j-th column, and $a_{i,j}$ be the output ($r \geq 0$). The new output of each element is computed as a function of the input:

$$a_{i,j} = \Phi(a_{i-1,j-r}, a_{i-1,j-r+1}, \dots, a_{i-1,j+r}),$$

where Φ stands for the processing rule. The parameter r measures the connectivity (or the cooperativity) of the system; a larger r means that a larger portion of the information contained in the previous processing layer determines the present state of each element, and that each element affects the state of a larger number of elements in subsequent layers. The *majority rule* seems to be the simplest rule for modelling the general properties of biological information processing [17]. According to this rule the state of an element is determined by the state of the majority of elements in the controlling sequence. Thus, if

$$\begin{aligned} A_i &= a_0 \cdots a_{n-1} \\ A_{i+1} &= b_0 \cdots b_{n-1} \\ \text{and} \quad A_{i+1} &= \Phi A_i \end{aligned}$$

then

$$b_j = \begin{cases} 1 & \text{for} \quad \sum_{\ell=-r}^{r} a_{j+\ell} > \theta \\ 0 & \text{for} \quad \sum_{\ell=-r}^{r} a_{j+\ell} < \theta, \quad 0 \leq j < n \end{cases} \tag{1}$$

where additions and subtractions on indices are modulo n, and θ is the threshold for the majority vote.

A network operating by the majority rule has a many/one as well as a one/many type of control, and an error-damping capacity, whose magnitude depends on θ and r. Owing to the latter property, many different initial configurations rapidly converge onto one *fixed point* , i.e., an n bits string that does not change under further operation of the function Φ. In our model the fixed point represents the *phenotype* of a given cell. The collection of configurations that lead to a given fixed point will be denoted the *domain of attraction* of this fixed point. A system with many fixed points can store a large amount of information, or, in biological terms, has a large phenotypic plasticity. A system with few fixed points has strong resilience, since many input errors are damped out without being expressed phenotypically.

2.1 Modulating mutation rate

It has been shown that the primary antibody repertoire is generated by somatic rearrangements of DNA segments in the course of B lymphocyte differentiation. Most of the elements from which the antigen binding sites are encoded are carried in the germline in sets of two or three. These are expressed in the antibodies in different combinations. A second molecular event that has been suggested as a mechanism for changing antibody expression is *somatic hyper-mutations*: the mutation rate at the immunoglobulin V region genes is estimated between $2 \times 10^{-2} - 4 \times 10^{-2}$ as compared with a figure less than 10^{-6} for a non-immunological lymphocyte gene. It has been further suggested that somatic mutations occur during the generation of immune memory and are responsible for tuning the response towards higher affinity [1,18].

The above description suggests that the antibody phenotypic repertoire is fully accounted for by accelerated genomic changes. This mechanism will be denoted the *Mutation Accelaration Mechanism* - *MAM*. We use our B-lymphocyte automata model for studying the effect of mutations on the phenotypic antibody repertoire. However, rather than limiting our investigation to genomic mutations we also check the effect of mutations, or more generally, errors in information transfer, in lower levels of the cellular organization. In particular we would like to know in what stage in the B-cell differentiation process do somatic mutations have the largest effect on the phenotype. To this end all possible 1-bit mutations are introduced at all the processing levels of all 2^n original model B-lymphocytes. Each of these lymphocytes is of information width, n, and of m processing levels; m being large enough so that, initially, the mth level information, describing the phenotype, is a fixed point. The information is processed from level 0 through m by the majority rule (Eq. 1). The threshold is taken to be $\theta = (2r + 1)/2$, and the connectivity is $r = 1$. It has been shown analytically that in this model, the 2^n genomes converge onto ϕ^n different fixed points, (or phenotypes), where ϕ is the golden ratio (for more details see 2.2).

The effect on the phenotype of each, 1-bit, mutation, is measured independently. First we calculate the proportion of cases in which a 1-bit mutation changes the phenotype, out of the total number of $2^n \times n \times m$ mutations. Results are presented in Table 1. Then the Hamming distance (HD) between the phenotype prior to every 1-bit mutation and the phenotype resulting from the mutation is calculated, and the average HD per phenotypic change is given in Table 1. In addition, this table shows the number of novel phenotypes added to the original repertoire. Note that at the present stage no special assumptions concerning antigen – antibody interactions are required and the model is a general model of cellular information processing. Assumptions that are unique to the antigen–antibody binding will be introduced into the model below.

It is clearly seen in Table 1 that mutations occurring at the genomic level have the largest probability of altering a single cell phenotype. Moreover, the size of the change, as measured by the HD between the newly generated phenotype and the original one is largest for mutations occurring at the genomic level. The effect of a single-bit mutation on the probability of altering the phenotype as well as on the average size of the change decreases with increasing distance from the genome of the mutating processing level. These results imply that in a committed homogeneous lymphocyte population mutations that occur at the genomic level should have the largest effect on the phenotype. In contrast, the effect of single-bit mutations on the number of phenotypes added to the original population repertoire is largest the closer the mutation level is to the phenotype. The reason for this seemingly contradiction is that in our model all the possible genomic configurations are already present in the population, so that genomic mutations can alter the phenotype of a single cell but do not add novel phenotypes to the original repertoire. Now, owing to its error damping

Table 1. Effect on phenotypic repertoire of all 1-bit mutations at different processing levels

n	Mutation level	Proportion of phenotype changes	Average HD under phenotype change	Number of added phenotypes
11	0	.77	2.16	0
	1	.64	1.42	22
	2	.59	1.26	22
	3	.57	1.23	22
	4	.57	1.16	154
9	0	.76	2.20	0
	1	.64	1.43	18
	2	.59	1.26	18
	3	.58	1.17	36
7	0	.72	2.30	0
	1	.62	1.40	14
	2	.58	1.20	14

capacity, the *majority rule* causes *contraction in phase space*, so that the number of possible configurations decreases with every additional level of processing. Thus in processing levels that are closer to the fixed point – the phenotype, the number of configurations generated by the processing of the genomic information is relatively small. Hence, mutations occurring at these levels have a larger probability of generating new configurations and, hence, of adding new phenotypes to the original repertoire. These latter results imply that errors in information transfer, occurring closer to the phenotype should be more effective in generating completely novel phenotypes.

2.2 Modulating the processing rules

A formula for the number of fixed points, $f(n, r, \theta)$, in networks of width n operating under the *majority rule*, with $\theta = (2r+1)/2$ and $r = 1$, is given in [19]. This function behaves asymptotically like ϕ^n, where $\phi = (1+\sqrt{5})/2$ is the golden ratio. It is easy to show that for $\theta = 2r$ only 2 fixed points exist: a homogeneous strings (000....0), whose domain of attraction is of size $2^n - 1$, and a homogeneous string (111...1), which is its own domain of attraction. A formula for the number of fixed points as a function of r, $f(n, r, \theta = (2r+1)/2)$, has been given and it has shown that the size of the phenotypic repertoire is a non-linear function of the network connectivity, r. The number of fixed points decreases with increasing r, whereas the stability of the phenotype to small (1 bit) input errors increases with r (AGUR, to appear). These results show that by modulating θ and r, the balance between variability (i.e., many fixed points with small domains of attraction), and resilience (i.e., a small number of fixed points with large domains of attraction) can be controlled. Thus, a cellular mechanism is postulated — the Processing RUles Modulation mechanism (PRUM), which operates by changing the connectivity among the elements of the cell, r, or by changing the error correction capacity of each element, i. e., the threshold, θ.

2.3 Comparing efficacy of different optimizers — PRUM vs. MAM

In the present chapter the postulated modulation mechanisms are compared for efficiency under the constraints of a trypanosome-like environment. Thus, the following modulation mechanisms are now introduced into our B-cell model.

PRUM Type-1, which under normal conditions operates by the majority rule, $\theta = (2r+1)/2$, $r = 1$; under stress (to be defined hereafter) it reduces the connectivity to $r = 0$; now the size of the controlling sequence of each element is 1, no error damping is possible and the phenotypic string is identical to the genomic string.

PRUM Type-2, which under normal conditions operates by the majority rule, $\theta = 2r$, $r = 1$; under stress it reduces the threshold so that now $\theta = (2r + 1)/2$.

MAM Type-3, which under normal conditions operates by the majority rule, $\theta = (2r+1)/2$, $r = 1$; under stress it accelerates genomic mutation rate, i.e., the probability of a single-bit alteration per cell replication, by a magnitude of α.

In these experiments the antigenic challenge (e.g., the dominant VSG) takes the shape of a n-bit string. This antigen is replaced by another one with some predetermined constant probability, P_f, with an additional parameter, A, being the maximal number of randomly chosen bits that can be altered every antigenic shift. Note that the probability that the same antigenic string will occur more than once, during the relatively short experiments presented here, is negligibly small.

The affinity of antigen–antibody binding is measured by the string complementarity. For example, an antigen represented by the string (111000000) is fully complementary with an antibody represented by the string (000111111). This *ideal phenotype* will have the largest proliferation rate. Reducing affinity, as measured by the HD from the *ideal phenotype*, reduces the rate of proliferation of the B-lymphocyte according to a predetermined function, which in these experiments is linear with the HD (Law = $Linear$). If binding affinity is equal to or smaller than a given threshold (here $Thr = 4$ out of a maximum of 9 possible complementary bits) then a stress signal is invoked, which turns on the cell specific stress modulation mechanism (Type-1, or Type-2, or Type-3). The modulation type is taken to be fully hereditary so that the offspring of cell Type-1, for example, will also be Type-1. In the experiments, whose results are described in Figs. 3, 4, the initial set of B-lymphocytes contains, $Pop = 100$, cells, each having a randomly chosen genome of size, $n = 9$, bits. This genomic information is processed in each of these lymphocytes through, $m = 5$, processing layers with layer 5 being the antibody. The binding efficacy of each antibody determines its proliferation rate, and since the population carrying capacity is, $K = 100$, cells, lymphocytes generating antibodies with higher binding affinity have a larger probability of being represented in the next generation.

The pathogen in this particular experiment shifts rapidly: the probability of antigenic shift is $P_f = .2$, and the maximal number of altered bits in the antigen is $A = 9$. The initial genomic mutation probability is the same for 0 to 1, and 1 to 0, bit alterations ($P_{01} = .01$,$P_{10} = .01$). The mutation acceleration rate for MAM (Type 3) is $\alpha = 10$. The duration of the experiment is $Gen = 100$. At the initiation of the experiment the population consists of the three cell types in roughly equal proportions. In the experiment whose results are shown Fig. 3 cells of type-3 (MAM) competitively exclude the two other cell types. In another experiment, having similar parameter values, whose results are shown in Fig. 4, PRUM cells of Type-1 competitively excludes the other cell types. That identical parameter sets can generate different experimental outcomes points out at the important role of stochastic factors in this system. Comparing both the histogram and the average population fitness curve in Fig. 3 and Fig. 4 it appears that PRUM Type-1 may have a faster recuperation rate following each

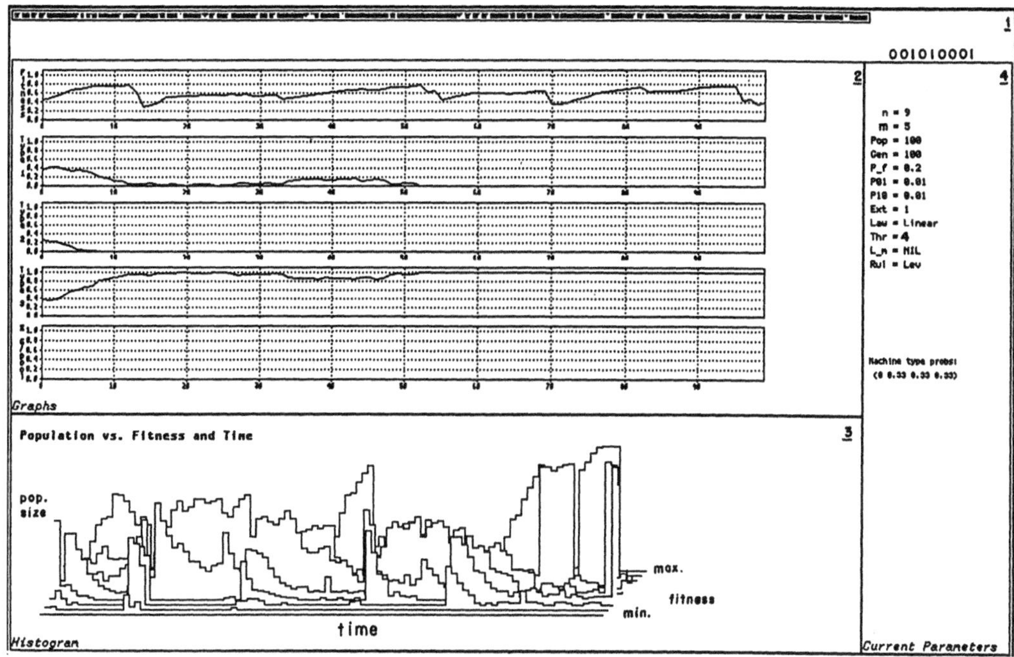

Fig. 3. A competition experiment between MAM (Type 3), PRUM (Type 1) and PRUM (Type 2). MAM competitively excludes the two other modulation types. Window 1 shows the exact picture of gene expression at time $t = 100$, for all 100 cells of dimensions 9×5; black and white pixels denote 1 and 0 respectively; the genomes are represented by the topmost layer while the phenotypes are represented by the bottom layer. The string at the lower-right side of the window is the *ideal phenotype* at time $t = 100$. Window 2 describes the graphs of the evolution of the process; *Fitness* – average population fitness in time; *Type 1–3*, proportion of cells of Type 1-3 in the total population in time. Window 3 displays temporal changes in the population fitness distribution. Window 4 displays major experiment variables (for details, see text).

antigenic change. In all experiments, except under one well-defined constraint (an antigen of the form 000...000, appearing shortly after the initiation of the experiment) PRUM cells Type -2 are rapidly excluded from the system.

Table 2 summarizes many competition experiments between MAM (Type-3) and each of the PRUMs separately, for different values of antigen shift probability and shift size. The initial proportion of MAM cells is .5 in all the experiments. Other parameters are as in Fig. 3. The final proportion of MAM averaged over 50 experiments for each set of parameters is given in the table.

The results presented in Table 2 show a clear disadvantage to the modulation mechanism PRUM Type-2, which is the strongest error damping mechanism of the three. This result is self-evident under the current constraints of the system, since the frquent changes in the antibody, forced by the frequent antigenic shifts of the pathogen, cannot be accommodated by this much too resilient mechanism. Our results in Table 2 do not show a clear advantage of any one of the two other modulation mechanism, even though a slight advantage of MAM is observable.

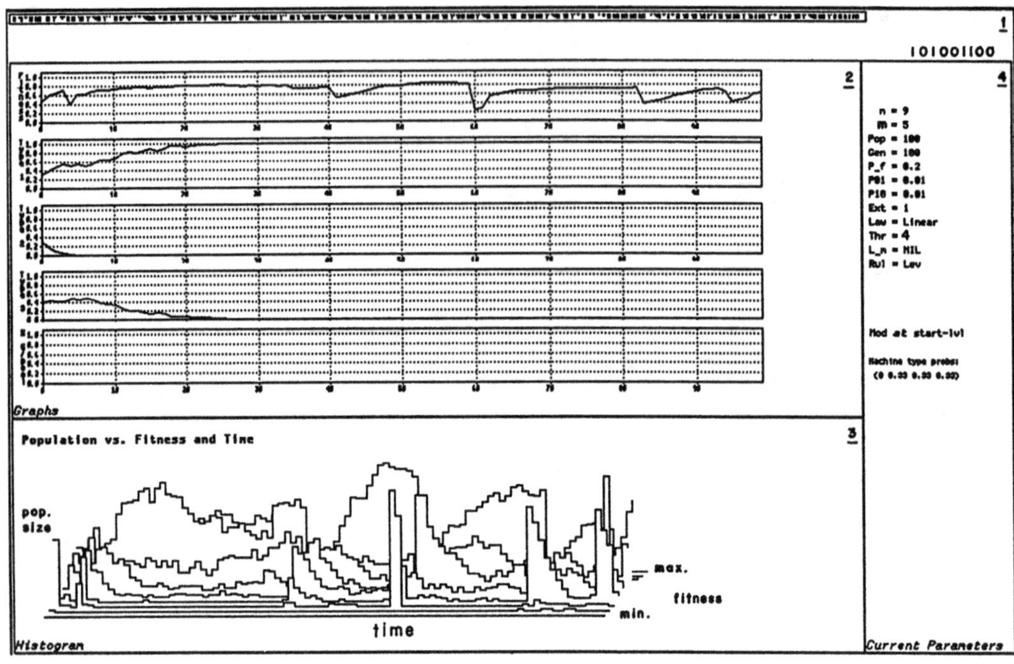

Fig. 4. A competition experiment between MAM (Type 3), PRUM (Type 1) and PRUM (Type 2). PRUM Type-1 competitively excludes the two other modulation types. Parameters as for Fig. 3.

Table 2. Competition results: MAM vs. PRUM Type-1 and MAM vs. PRUM Type-2

Max. bit changes per antigenic shift	Probability of antigenic shift	Final proportion of PRUM type 1	Final proportion of PRUM type 2
1	.05	.46	.16
	.20	.50	.10
4	.05	.47	.10
	.20	.34	.26

2.4 Conclusions

Immunologists maintain that the antibody repertoire is primarily determined by genomic events: rearrangements and hyper-mutations. Somatic hyper-mutations are suggested to be part of the memory pathway, where a new repertoire of binding sites with high affinity for the antigen is generated for increasing the efficacy in controlling subsequent infection by an identical antigen. Results of the present work only partly support this view.

Our results show that genomic mutations are effective in altering single cells' phenotypes. In contrast, mutations occurring at lower levels of the cellular genomic information processing should be more effective when a momentary expansion of the whole phenotypic repertoire is required.

Previous competition experiments, published elsewhere, were constructed so that the "ideal phenotype" fluctuated between two largely different states only [17]. That model, representing alternating periodic infections of two distinct antigens, showed a clear advantage of PRUM Type-1 on MAM. As the present model does not reflect a similar PRUM superiority, the difference between the two models may be of interest. A possible explanation for this difference may be that PRUM is a more effective optimizer when preserving the memory of previous infections is advantageous. In contrast, the experiments brought forward here are constructed so that the probability of re-emergence of the same antigen is negligibly small. Hence, the memory of previous antigen—antibody encounters does not contribute to the response efficiency. Whether this is the reason for the failure of the PRUM in outcompeting the MAM is still under investigation. Results will be reported elsewhere.

The model presented in this work enables formal verification of the efficacy of postulated optimization mechanisms in the immune system. This model (written in LISP) is much more flexible in its assumptions and, hence, much more realistic, than more conventional mathematical descriptions of this system. Further elaboration of the model is warranted in order to elucidate the molecular mechanisms underlying the immune response to parasitic invasion.

References

1. I. Roitt: *Essential Immunology*, Blackwell Scientific Publications (1988).
2. D.G. Bounds: Nature **329**, 215 (1987).
3. K.J. Vickerman: Cell Sci. **5**, 163 (1969).
4. K.J. Vickerman: Nature **273**, 613 (1987).
5. G.A.M. Cross: Parasitology **71**, 393 (1975).
6. P. Borst: Ann. Rev. Biochem. **55**, 701 (1986).
7. Capbern: *et al.* Parasitol. **42**, 6 (1982).
8. L.H.T. Van der Ploeg, D. Valerio, T. De Lange, A. Bernards, P. Borst, F.G. Grosveld: Nucl. Acids Res. **10**, 5905 (1982).
9. K. Vickerman & J.D. Barry: In *The Immunology of Parasite Infection* ed. by S. Cohen & K.S. Warren, Blackwell Scientific Publications, Oxford (1982).
10. R. Ross & D. Thomson: Proc. R. Soc. Lond. Biol. **82**, 411 (1910).
11. K.J. Vickerman: In *Parasites in the Immunized Host: Mechanisms of Survival*, Ciba Foundations Symposium (Elsevier), pp. 53–80 (1974).
12. J.E. Donelson, M.J. Turner: Sci Amer. **252**, 32 (1985).
13. T. Baltz, C. Giroud, D. Baltz, C. Roth, A. Raibaud, & H. Eisen: Nature **319**, 602 (1986).
14. A.W.C.A. Cornelissen, P. Johnson, J.M. Kooter, L.H.T. Van der Ploeg & P. Borst: Cell **41**, 825 (1985).
15. E.N. Miller & M.J. Turner: Parasitology **82**, 63 (1981).
16. D. Abiri, L.H.T. Van der Ploeg & Z. Agur: (to appear).
17. Z. Agur: IMA Jour. Math. App. Med. Biol. **4**, 295 (1987).
18. K. Rajewski, I. Förster, A. Cumano: Science **238**, 1088 (1987).
19. Z. Agur, A.S. Fraenkel, S.T. Klein: Disc. Math. **70**, 295 (1988).

Index of Contributors

GPSR Compliance
The European Union's (EU) General Product Safety Regulation (GPSR) is a set of rules that requires consumer products to be safe and our obligations to ensure this.

If you have any concerns about our products, you can contact us on

ProductSafety@springernature.com

In case Publisher is established outside the EU, the EU authorized representative is:

Springer Nature Customer Service Center GmbH
Europaplatz 3
69115 Heidelberg, Germany

www.ingramcontent.com/pod-product-compliance
Lightning Source LLC
LaVergne TN
LVHW080848170826
845678LV00006B/1744
9783642839375